TOP 10 DESIGNERS RECORD

名师录 设计界十人

深圳市智美精品文化传播有限公司 编　刘小鹏 译

大连理工大学出版社

图书在版编目 (CIP) 数据

名师录：设计界十人：汉英对照 / 深圳市智美精品文化传播有限公司编；刘小鹏译. —大连：大连理工大学出版社，2013.4
ISBN 978-7-5611-7738-9

Ⅰ. ①名… Ⅱ. ①深… ②刘… Ⅲ. ①室内装饰设计－作品集－中国－现代－汉、英 Ⅳ. ① TU238

中国版本图书馆 CIP 数据核字 (2013) 第 053038 号

出版发行：大连理工大学出版社
（地址：大连市软件园路 80 号 邮编：116023）
印　　刷：利丰雅高印刷（深圳）有限公司
幅面尺寸：245mm × 320mm
印　　张：20
插　　页：4
出版时间：2013 年 4 月第 1 版
印刷时间：2013 年 4 月第 1 次印刷
策划编辑：袁　斌　刘　蓉
责任编辑：刘　蓉
责任校对：李　雪
封面设计：张　群

ISBN 978-7-5611-7738-9
定　　价：320.00 元

电话：0411-84708842
传真：0411-84701466
邮购：0411-84703636
E-mail:designbooks_dutp@yahoo.com.cn
URL:http://www.dutp.cn

如有质量问题请联系出版中心：（0411）84709246 84709043

Whenever there are new books published, I always cannot wait to read. On the one side, I want to know the change in the design trend, and on the other side, I want to learn from others the design thinking and logic, as well as other creative ideas.

It has been more than a decade since I came into this industry. Honestly speaking, my mind is kind of elastic fatigue, feeling that I am confined to a design frame. If there was still no new stimulation, I am afraid that my motivation and drive for creation will be on the slide.

Today, the design publishing industry is enjoying a rapid development, especially in recent years, it has undergone tremendous changes. Several times as many reference books as before have emerged in the market. Sometimes, reading even seems to be a burden as it takes quite a long time simply to pick up good and appropriate books, and time is just what I'm lack of most.

If each design work is a dedicated one of the designer, then each design book is what editors have racked their brains to select, compile and evaluate carefully from these dedicated works; therefore, such mental and physical efforts deserve heartfelt respect. In such design environment, the latest design methods of top masters are the mainstream of the design industry, which lead the direction of the design development. What is included in this book is the latest tour de force from ten design masters. I sincerely hope that this book will be a timing gourd of fresh water for all in the design industry on their lonely design road, which can offer professional guidance for those who have interest in design, and provide more inspiration to the designers in positioning their design direction. Only in this way, can the social responsibilities of publishers be brought into full play.

For mutual encouragement

Hank M.Chao

每每有出版社出新书，我总是迫不及待地想看看。一方面是想了解设计趋势上的变化，另一方面是想从中学习别人的设计思维和逻辑，还有其他创新的想法。

入行已经十多年了，说实话，脑袋真的有点弹性疲乏，总觉得老是在一个设计框构里边跳不出来，估计如果没有新的刺激，创新的动机和动力将会每况愈下。

现在，中国的设计出版业正在蓬勃发展，特别是近几年更是有了巨大的转变。市面上的参考书籍比以前多了好几倍，有时候连看书都觉得是一种负担了，因为单是从中挑选好的适合的书都会花很长的时间，而时间恰恰又是我最缺乏的东西。

如果说每个作品都是设计师的呕心沥血之作，那么每一本设计书籍则都是编辑从这些呕心沥血之作里面再绞尽脑汁挑选、编辑和认真评价的过程，这样的努力和心力是让人打从心里敬佩的。当前设计环境下，名师们的最新设计手法是设计界的主流，主导着设计发展的方向，而本书收录的也正是当下设计界10位名师的最新力作。由衷地希望本书能成为设计界的芸芸众生在孤独的设计路途上一瓢适时的清水，给对设计有兴趣的朋友一些专业的引导，为设计师在策划设计方面增加一些灵感。如此，才是充分发挥了出版业者的社会责任。

以此共勉

赵牧桓

Interview with Designer Yellow
陈武设计师专访

1. There is a clear distinction between public decoration and domestic decoration in interior deign. Would you please share you views on the biggest difference between them and what we can do to integrate them?

Yellow: They are two different patterns. Public decoration needs to cater to the esthetic taste of the public as well as the functions of the business while domestic decoration only needs to appeal to the owner's personal taste of life. What the difference is when you do the public designs, you should pretty clear about the business operation and business pattern so that you can pick out the common elements, while for the domestic ones, you only have to highlight the individuality in the project.

2. What do you think is the most important consideration in the design of business space? What do we do to make it happen?

Yellow: The basic considerations are the function of the business, the return on investment and the common esthetic taste of consuming groups. You need to put yourself in the shoes of investment party and operating party. You need to learn and to experience how to run a place before decorating it. In a way, the designer acts as the business consultant. A business design which does not profit can never be a good design. Therefore, I believe: "there is no best design, only the right design."

3. The decoration of night clubs (bars, KTVs etc.) can easily be classy as well as be tacky. What miracle work do you do to integrate both?

Yellow: The key is to weaken the design to such an extent that it becomes objective experience. Most consumers are nonprofessionals. They want friendly, interactive and harmonious environment instead of a gallery. This policy applies to other business spaces other than night clubs. It's important to maintain the theme of the decoration but the customer feelings are equally significant.

4. At present, the design pays more and more attention to soft decoration. In your opinion, how is it applied to the public decoration project?

Yellow: The application of soft decoration and furniture is the key of modern decoration. A lot of domestic decoration elements are employed in modern business space decoration. They bring affinity and warmth to these places.

5. Lighting is a significant part in business projects. How do you make the best use of it to achieve the desired effect?

Yellow: Lighting requires very professional skills in interior decoration. The decoration businesses abroad usually own their professional team to work on it. In China, there are teams like these, too. For our big projects, it is necessary to hire the professional team to do the further design.

6. Your projects have various features. What elements are involved to determine the styles?

Yellow: The elements which can determine my styles are: the orientation of the project; the needs of the clients and the distinctive cultural orientation.

7. Would you please share your insights about the future development of this business?

Yellow: The trend will be multi-development in the way of industrial chain. We'll try to turn the original concepts into reality. In the meantime, we'll develop Xinye furniture and soft decoration products, offer design service of brand VI. We'll keep pushing the boundaries, try different styles and elements and develop original products. We will attend to every detail, for they determine the quality and success of the project..

1. 室内设计中，公装和家装设计群分割较为明显，请您谈谈公装项目与家装项目在设计上最大的区别在哪里？如何形成共融？

陈武：两种业态不同，公装要满足商业所有的大众审美和功能需求，而家装只需要满足小业主的个人生活审美需求。区别在于做公装设计，必须要懂得商业运作和业态功能，这样才能准确地把握共性，家装则只要体现业主个性即可。

2. 您认为商业空间的设计最注重什么？怎么把握这个重点？

陈武：最基础的是要了解业态的功能需要、投资回报及消费群体的审美共性。要站在投资方和运营方的角度去考虑设计，去学习和体验如何经营，再来设计。某种意义上设计师充当了商业顾问的角色，不赚钱的商业设计不是好设计，所以我认为：“没有最好的设计，只有最合适的设计。”

3. 夜店（酒吧、KTV 等）项目可雅可俗，那么在您的作品中，是怎样做到雅俗共赏的？

陈武：把设计感弱化到客观体验中是关键。消费者大部分是非专业人士，他们需要亲和、互动、和谐的环境，而不是美术馆。这个要求不光是针对夜店，其他商业空间也同样适用。在保留主题创意方向的同时，客户感受也很重要。

4. 如今的设计越来越重视软装的使用，在您看来，工装项目中的软装该怎样来搭配？

陈武：软装及家具的使用是现代装饰空间中最核心的部分。现在的商业空间中运用了大量的家装元素，大大增强了商业空间中的亲和力和温暖感。

5. 商业项目中灯光和照明是很重要的一环，您是怎样运用它来达到想要的效果的？

陈武：灯光设计是室内环境设计中非常专业的一环，在国外的同行业中，都有专业的团队来做配套设计。现在国内也有专业团队专攻灯光设计，我们现在的大型项目，也是要请专业团队配合做深化设计的。

6. 您的项目大多各有特色，风格定位是怎样决定的？取决元素有哪些？

陈武：要根据项目的定位、终端客户的需求和差异化的文化导向来决定我对风格元素的取向。

7. 您对未来的设计之路有什么样的看法或想法？

陈武：我们会加更多元化地发展，以产业链的方式发展业态，让更多更好的创意落地；开发新冶家具和软装的产品，增加品牌 VI 设计服务等。我们会不断超越自己，尝试不同的风格和元素，并大胆开展创意产品的研发工作，让每个细节来决定项目的品质和成败。

1. What is the major difference between democratic decoration projects and public ones? Where do you put their emphasis on?

Jianguo Xu: The major difference between the two lies in the different target customers that they serve. The target users of domestic decoration are a combination based on households, whereas public decoration often meet and serve social public and operators. From the perspective of central focus, the domestic decoration deals more with people's daily life, focusing on making people know more about life, while the public decoration tends to focus on improving the quality of life towards space. Simply speaking, one is built-in, and the other is presentational.

2. What is your design concept of Beijing Shouzhou Hotel? Are there any cultural settings concerning your choice of style and theme? Could you please talk in detail about your design process of this project?

Jianguo Xu: Beijing is an international metropolis that integrates politics, economy and culture into one; whereas Shouzhou is simply an old city with a long history. Therefore, when it comes to the choice of style, we want it to possess the feel of ancient city Shouzhou, and at the same time, we hope it can take on an internationalized appeal.

After we got the project, we went to its site and spent nearly three months to research the local culture and common folk-custom. I realize that although it is an old city, it is old only in appearance; inside it is actually a very modern city. As a result, we want to present a sense of "impression".

Hence, in the process of design, we went through the early and the middle stages before we got the final draft in the late stage. At last, Beijing Shouzhou Hotel is not only an embodiment of Shouzhou culture, but also a small step we take towards the internationalization of Chinese culture. Hopefully it is a direction, and can bring a tiny inspiration to us folks who do design now.

3. What is the most important element in Chinese-style design in your opinion? How to convey the subtle enchantment of Chinese style and not to make people feel suppressed at the same time?

Jianguo Xu: I think what counts most is the "feel", and there are many ways and methods in design. Traditionally, Chinese people who did design were poets. Consequently, they concerned more about the proportion of the measuring scale and the elegance of lines. So design depends on "feel" most of the time.

4. Usually, Chinese style can convey a sense of Zen. As far as you are concerned, how can this sort of invisible things be conveyed?

Jianguo Xu: I think it needs us to feel and to sense. In my view, Zen is only one aspect, and to be more specific, it is about showing the nature of humans, making people feel more naturally. So it is still a kind of "feel".

5. How do you look at the blend and integration of culture and environment? And how do you show it in your works?

Jianguo Xu: I think the blend and integration of culture and environment is of extreme urgency right now. As far as my works are concerned, I would pay a close attention to originality; to refine the environmental and cultural elements, and hope the occupant can feel my endeavor; and understand it so that it can be interpret into design language for more people to reach it and make it more exquisite, more civil and better blending into the environment. As a consequence, it is not only for culture and environment to blend and integrate, but also to be linked closely together, and both of them are indispensable.

6. There is an increasing focus on the use of soft decoration in today's design, as far as you are concerned, what are the differences in the application of soft decoration in public projects and domestic ones?

Jianguo Xu: In the application of domestic decoration, the soft decoration is for a real use, but in public decoration, it is for display most of the time, a way of demonstration. Domestic decoration should not be defined as show flats, and should lay its focus on life. You should like it, otherwise you would not use it. However, what pubiic projects pursues is mainly perfectionism.

7. No matter business projects or domestic ones, lights and illumination are really important. So in your eyes, what are the design differences of the two? And how do you employ them to reach your desired effect?

Jianguo Xu: Lights and illumination are important. Without lighting, our design cannot "light it up" at night. It is also a vital part of a design, the core of design. In terms of business and domestic projects, I feel that more difference lies in the patterns of manifestations of lights, domestic decoration may put its emphasis on the basic lighting and the illumination of the environment, and use as little as possible the indirect lighting; while business lighting is always pursuing the extreme lighting, and that is the full sense of "lighting-up".

1. 您觉得家装项目和公装项目在设计上最大的区别在哪里？各自侧重什么？

许建国：家装与公装最大的区别在于其所面对的使用人群的不同。家装面临的人群是一个以家庭为单位的组合，公装面对和服务的往往是社会大众和经营者。从侧重点方面来说，家装设计更多地是解决人的日常生活，让你更懂得生活；而公装往往会提升我们对于空间生活品质的要求；简单来说，一个是内置的，一个是表象的。

2. 北京寿州大饭店的设计思路是怎样的？风格与主题的选择有什么文化背景吗？能具体说说该项目的设计历程吗？

许建国：北京是一个集政治、经济、文化于一体的国际化大都市，而寿州却只是一座有着很长历史的古城。因此，我们在风格的选择上，希望它既具有古寿州的感觉又能表现出一种国际化的风格。

接到案子后我们去了项目所在地，用了将近三个月的时间对当地的文化、基本的民俗进行了采风。我发现寿州虽然是古城，但这只是一个很表象的外衣，里面其实是一个很现代的城市。所以，我们希望能做出一种“印象”。

因此，在设计过程中我们经历了前期、中期和后期的最终定稿。最终，北京寿州大饭店不仅仅是一个独立的寿州文化，也是我们对中国文化走向国际化所做的一个小小的努力。希望它是一个方向，更多的时候是希望它能给我们现在做设计的这帮人一点小小的启发。

3. 您觉得中式风格设计中最重要的因素是什么？怎样才能既传达出中式的意蕴美，又不会浓烈得让人觉得压抑？

许建国：我觉得最重要的还是“感觉”，其实设计有更多的方式与方法。中国人自古以来一直是一帮诗人在做设计，所以很讲究比例尺及线条的优美，所以更多的时候是凭着“感觉”来的。

4. 中式风格往往能传达出一种禅意的境界，在您看来，这种无形的东西要怎样来传达呢？

许建国：我觉得还是需要我们去感受、去感知。我觉得禅意只是一个方面，更确切的说是表现人的人性这一方面，让人感觉更自然，其实也还是一种“感觉”。

5. 您怎样看待文化与环境的交流与融合？在您的作品中又是怎样来表现的？

许建国：在当下，我觉得文化与环境的交流与融合是一件迫在眉睫的事情。对于我的作品来说，我会很关注原创性；把环境和文化提炼出来，希望业主方能看出我的付出，通过了解读懂其设计语言，让大众更多地去接触它，使它变得更精髓、更文化、更能融入环境，所以文化与环境不光是交流与融合，而应该是紧紧相连，缺一不可的。

6. 如今的设计越来越重视软装的使用，在您看来，公装项目与家装项目中软装搭配的运用有什么不同？

许建国：家装的使用中，软装是真正拿来使用的，而公装应该大部分是一种展示的作用，一种表现的方法。家装不应该定义为样板间，对于家装还是要注重生活，你必须喜欢它，不喜欢你不能拿来用；而公装项目主要还是追求完美主义。

7. 不管是商业项目还是家装项目，灯光和照明都很重要，那么在您看来，这二者的设计有什么不同，您又是怎样运用它们来达到想要的效果的？

许建国：灯光和照明都很重要，如果离开了灯光，我们的设计在夜晚是无法“点亮”的。它也是设计最关键的一部分，是设计的核心。对于商业与家装项目，我觉得更多的还是灯光的表现形式有所不同，家装可能注重的是基本照明与环境灯光，间接灯光尽量少用；商业照明大家一直还是追求极度照明，也就是“点亮主义”。

Interview with Designer Jianguo Xu
许建国设计师专访

YELLOW 陈武

RAYNON CHIU 邱春瑞

DANFU LIU 刘卫军

SHUHENG HUANG 黄书恒

RICKY WONG 黄志达

HANK M.CHAO 赵牧桓

FRANK JIANG 姜峰

ERIC TAI 戴勇

JIANGUO XU 许建国

DORIS CHUI 徐少娴

YELLOW
陈　武

深圳市新冶组设计顾问有限公司 负责人

新冶（香港）设计工程有限公司 董事
中国区"夜时尚"大型跨界联盟活动 发起人
中华民族文化促进会 会员
北京欢乐时空动漫学院 客座教授
国际室内建筑师 & 设计师理事会（ICIAD）理事
国际室内设计师 & 室内建筑师联盟协会 会员
中国室内装饰协会 会员
IFI 国际室内设计师协会 会员
深圳市室内设计师协会 会员
当代中国设计与价值合作事业联盟（UCDVR）专业设计顾问

获奖经历：

多次荣获国内外设计大奖，近期凭借重庆夜色酒吧获"金指环 2009 全球室内设计大奖"金奖。

设计理念：

率领一支狂热追逐现代时尚艺术的创意智囊团队，崇尚简约原创、幽默纯熟的设计理念，追求宽泛的视觉语言。以敏锐的触角，解读时尚密码，坚持"策略为中心，创意为灵魂"的指导思想，构建了一个兼具国际视野、立足本土，沟通中国市场与文化气脉的经营团体，所承建项目遍布全国各地。

凭借对西方异域文化及国际时尚理念的深切领悟和独到见解，善于将时尚、艺术、科技、文化完美结合，成就了众多东西方文化混搭的经典之作。

致力于将新冶组打造成为中国最具时尚文化的空间设计典范，以构想客户共鸣为创作意念，倾力为客户打造饱含独特气质并具时尚创意的空间设计方案和整体优质的配套服务！

目前，新冶组分公司脉络已延伸到上海、重庆、杭州等一线城市……

项目地址
长沙

面积
554 平方米

设计师
陈武

设计公司
深圳市新冶组设计顾问有限公司

主要用材
黑色光面石材、光面大花白石材、毛面水纹银石材、实木仿古木地板、灰麻石踏板、灰镜、黑色镜钢、金属挂网、白色人造石吧台面、真空隔音玻璃

DONE CLUB

CHIVAS
敢先锋 不独行
芝华士 J&J创始纪念版
CHIVAS

Done Club is located in the central area of Changsha City, created for the emerging noble and trendsetters in the city. Entering the bar lobby from the busy streets, you can see the specially treated LOGO wall made of rotten copper grill, without repeating the vulgar posture of bars in the past, showing an elegant and artistic atmosphere; the stone is cut into specific shapes to form floor and walls, with expensive cost, creating the space quality in silence; the glass wall on the left and the inkjet painting bring customers a gorgeous sense.

The main hall of the bar makes use of a large number of gray mirror, black steel mirror and other reflective materials, coupled with lighting effects of different colors to create a romantic atmosphere; in the atrium, the pentagon shaped tables look like divided cells, forming seamless combination according to the customer flow and their requirements, and such arrangement not only meets the commercial demand of the bar business, but also leaves social networking space for customers.

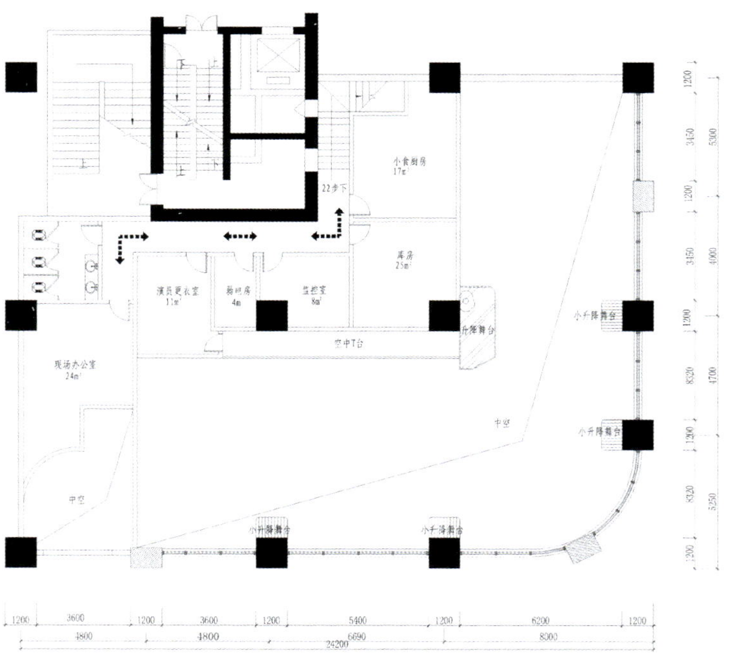

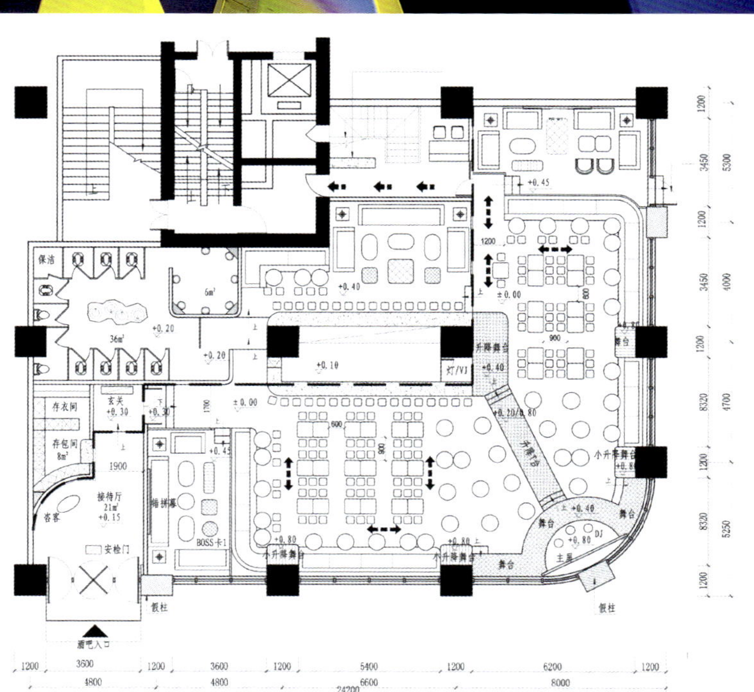

DONE CLUB 位于长沙市的中心区域，专为城中新贵、潮人打造。从繁华的街道进入酒吧前厅，经过特殊处理的腐铜格栅 LOGO 墙，不复以往酒吧的庸俗之姿，显出雅致、艺术的气息；将石材切割成特定造型拼接的地面与墙身，成本昂贵，于无声处塑造出空间品质；左面的玻璃墙与喷绘彩画带来华美之感。

酒吧主厅运用了大量的灰镜、黑色钢镜等反光材质，搭配不同色调的灯光效果，营造出浪漫的氛围；中庭里五边形造型的酒桌犹如一粒粒分裂的细胞，可以根据客流量及顾客的要求进行自由无缝式组合，此安排既满足了酒吧经营的商业需求，又给顾客预留了社交网络空间。

International

Pioneer

D-ONE CLUB
夜生活名片
时尚 品質

消防栓
V188

Chengdu 1982 Red Wine's Living Museum

项目地址
成都

面积
796 平方米

设计师
陈武

设计公司
深圳市新冶组设计顾问有限公司

主要用材
做旧橡木地板、镂空雕刻板、做旧铜板、香槟金银箔、做旧红砖、金属网帘

成都 1982 红酒音乐生活馆

The side wall of the building is set with "1982", avoiding direct publicity, and the LOGO appears in the east with bold style and concise lines to highlight the design intent of "meeting dawn intimately in the first time" and the profound meaning of praying for the brand. The designer spreads elegance in every corner of the space, showing the post modern European flavor, and highlighting the unique trajectory of spirit with refined layout.

The 1982 is a living and tasting museum integrating with red wine, private parties, art show and butler service to form a one stop system. The 3D-styled space line at the entrance is clean yet plump, and the champagne gold beam columns connect the ceiling and the floor of duplex building, where the right and left beam columns circle around to the ceiling to pave an "o" shape with edges linked to form a ceiling totem in the shape of "8", finally becoming the visual focus of the space.

The designer cleverly implants the brand character to give customers a feeling of "concise but not simple". The design of antique oak floor, well-proportioned wooden red wine buckets, champagne metal birdcage and lighting reflects the brand connotation which is rustic, simple and highly private.

82
Red wine's
Living Museum
CLUB
LOVE
FREE CLUB
Red wine's
LOVE LIFE
BAR
Living Museum
BEAUTIFUL
PERSISTENY
LOVE IS A TENDER LOOK
LOVE IS AN ART
YOU ARE SO BEAUTIFUL
Frank Phélan
Saint-Estèphe

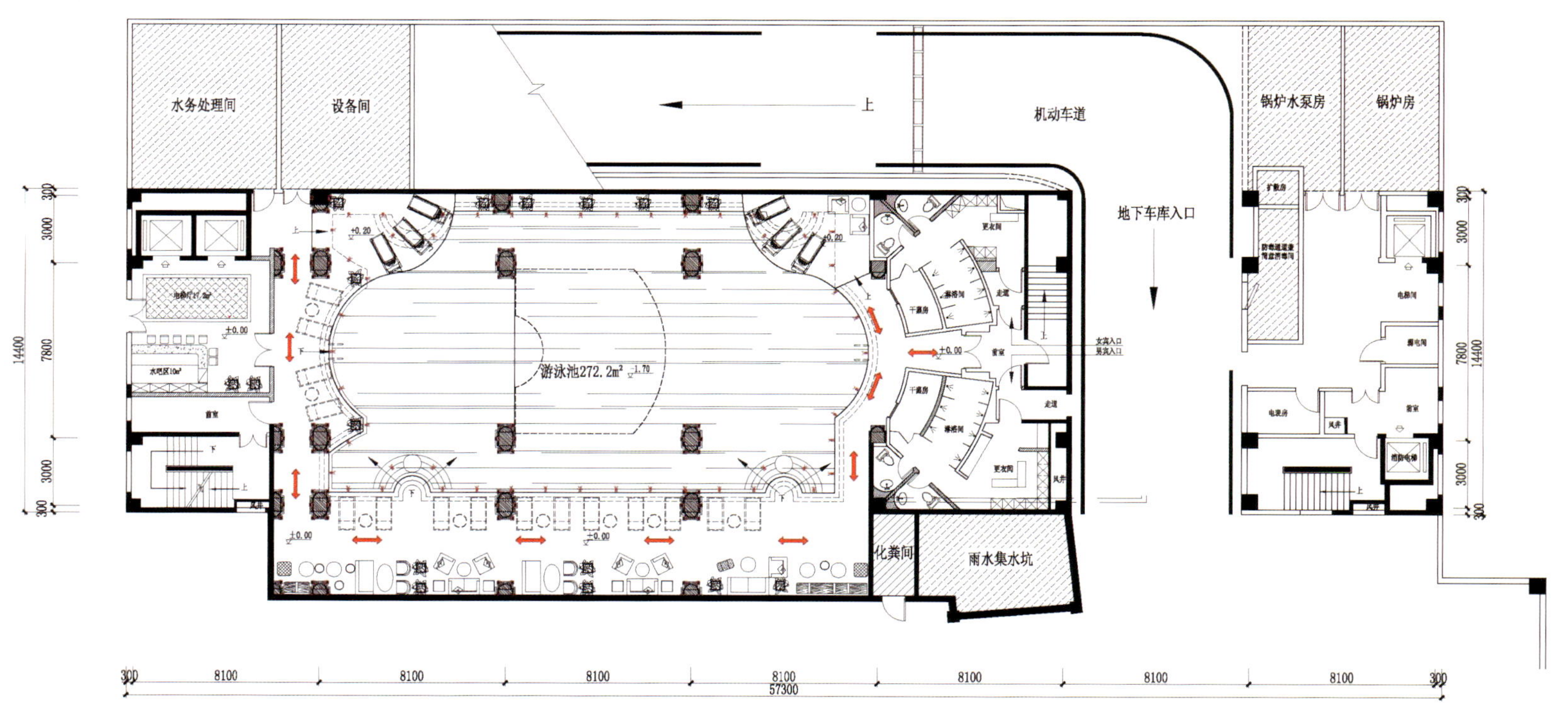
水务处理间
设备间
上
机动车道
锅炉水泵房
锅炉房
地下车库入口
游泳池272.2m²
化粪间
雨水集水坑
300
8100
8100
8100
8100
8100
8100
8100
300
57300
14400
3000
7800
3000
300

CLUB
YOU ARE SO BEAUTIFUL
FREE CLUB
BAR
LOVE
Red wine's
LOVE LIFE
FREE Living Museum
Red wine's
Museum CLUB
Living Museum
BEAUTIFUL LOVE PERSISTENY
LOVE IS A TENDER LOOK
PERSISTENY
Red wine's
WHICH BECOMES AN ART
BEAUTIFUL
LOVE LIFE CLUB
FREE
Frank Phelan

BEAUTIFUL
LOVE LIFE CLUB
FREE
Red wine's

PARIS
SAINT-EMILION GRAND CRU
Arômes de Pavie
Réserve
Château Croix-Mouton
CHATEAU DAUZAC
MARGAUX
Frank Phelan

建筑侧墙体镶嵌着“1982”，避开正面的张扬，LOGO 出现在东侧，造型大胆、线条简洁，凸显了“与晨曦第一时间亲密相见”的设计意图与祈愿品牌的深刻寓意。设计师将优雅植入空间的每一个角落，绽放后现代欧陆情调，以脱俗的布局强调与众不同的精神轨迹。

1982 是集红酒、私人派对、艺术秀、管家服务体系于一体的生活品鉴馆。在大门口看，3D 式的线条空间干净而不失丰腴，香槟金梁柱驳接天花与复式楼地板，左右梁柱盘旋至天花平铺成“O”形，边缘相接，形成“8”字形天花图腾，成为视觉焦点。

设计师巧妙地植入了品牌字符，给人以“简约却不简单”的感受。做旧橡木地板、错落有致的木质红酒桶、香槟色金属鸟笼、灯饰的设计运用，无不体现出质朴、简约、高私密的品牌内涵。

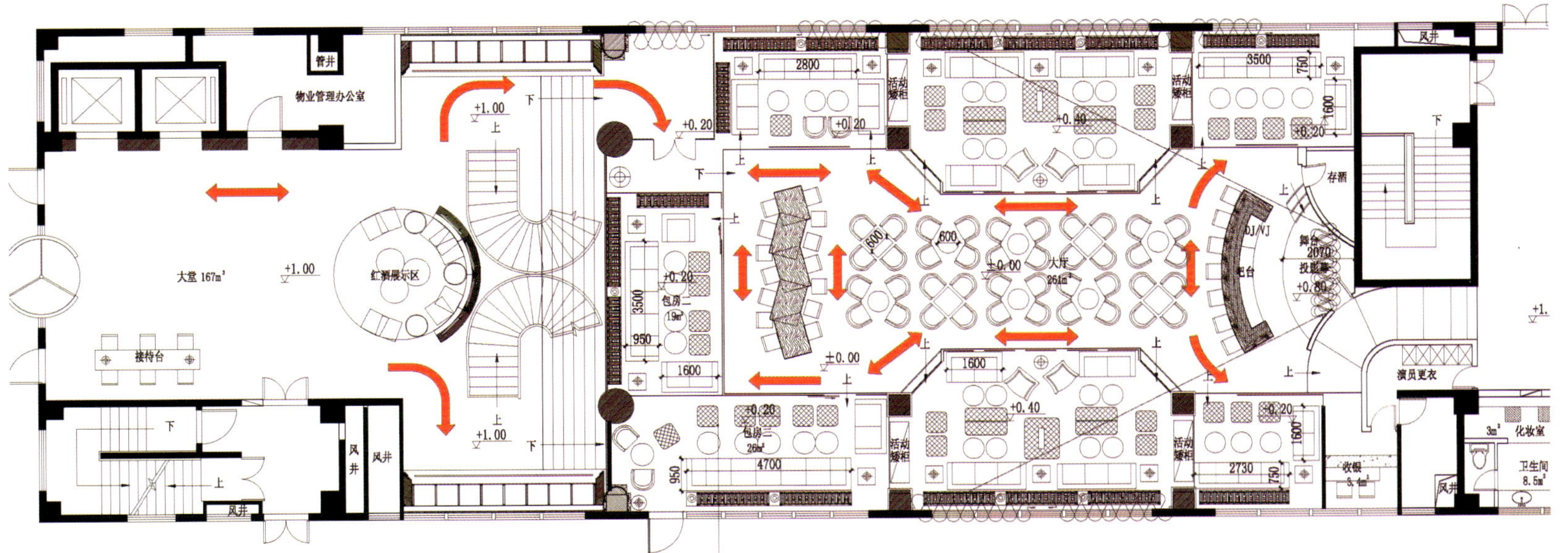

管井
物业管理办公室
大堂 167m²
红酒展示区
接待台
风井
包房一
19m²
包房二
26m²
活动矮柜
大厅
261m²
DJ/VJ
吧台
舞台
投影幕
存酒
演员更衣
化妆室
卫生间
8.5m²
收银

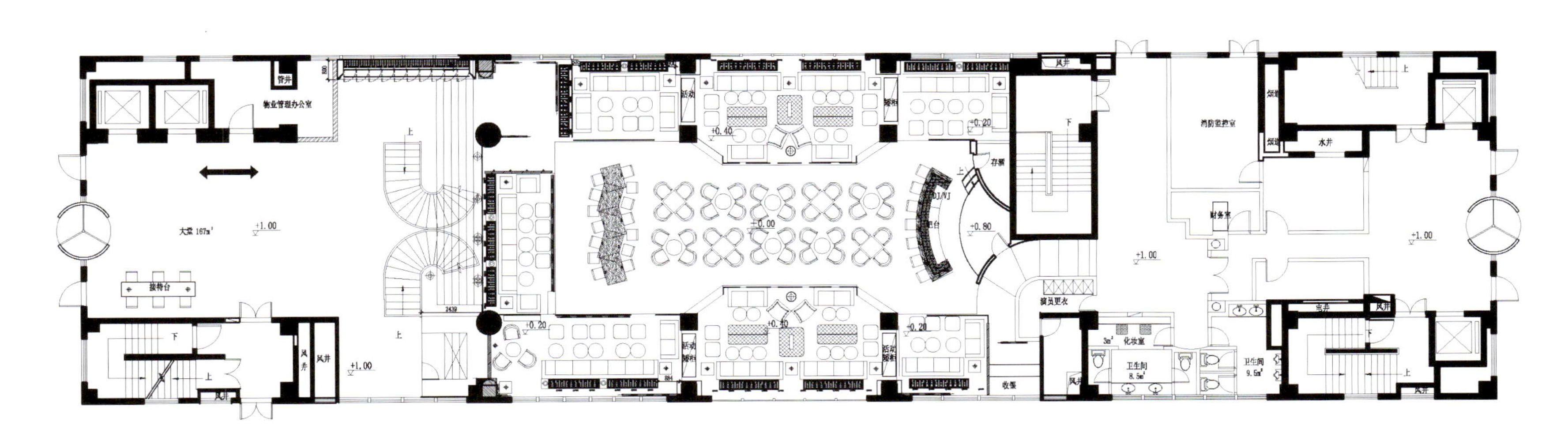

管井
物业管理办公室
大堂 167m²
+1.00
接待台
下
上
风井
+0.40
+0.20
+0.00
2439
884
DJ/VJ
存酒
+0.80
演员更衣
化妆室
卫生间
8.5m²
9.5m²
收银
消防监控室
水井
财务室

Goodview Hotel Colors Blaze Entertainment Plaza

项目地址
东莞樟木头

面积
5500 平方米

设计师
陈武

设计公司
深圳市新冶祖设计顾问有限公司

施工单位
东莞市海力峰装饰有限公司

三正半山酒店万紫千红娱乐广场

The design is expanded around the element "water", where the entertainment plaza is divided into leisure area and dynamic area, equipped with a huge water playground, and contains various experiential entertainment projects. In the leisure area, the colorful background lights with great momentum are used to create a dream of flowers falling. In order to achieve the intimate interaction between the entertainment projects and customers, the dynamic area introduces a variety of high-tech techniques, where the glass tubes are filled with water to create a dreamy scene of beauty dancing, and the hologram technology shows a nearly real effect of six-dimensional space.

The DJ platform is floating in the center of the water pool, where in the pool, "mermaids" service you with good wine, people dance in the air for you, and the spider-men may climb to your private booth. Traditional stage performances, such as acrobatics, dance and drama, are transformed into interactive entertainment projects, here the design breaks the limits of space, changing into an artistic operation of the whole entertainment projects.

设计围绕“水”元素展开，娱乐广场分为休闲区和动感区，设有巨大的水秀场，各种体验式娱乐项目也从中演化出来。休闲区用气势恢弘的背景花灯营造出落英缤纷的梦境。动感区为了取得娱乐项目与顾客之间的亲密互动，采用多种高科技手法，玻璃管内注满水营造出美人起舞的梦幻场景，全息影像技术呈现出逼真的六维空间真人效果。

DJ台起伏在水池中央，池中“美人鱼”为你端上美酒，空中飞人为你翩翩起舞，还有蜘蛛侠爬上你的包厢。传统的舞台表演项目如杂技、歌舞、戏剧等都演化为互动娱乐项目，设计在这里突破了空间局限，上升为整体娱乐项目的艺术运作。

C9

Lone I Club

项目地址
深圳

设计师
陈武

设计公司
深圳市新冶组设计顾问有限公司

主要用材
实木板刻效果、铜板墙面、金属链天花、沉船木地板

泷 爱

The Chinese name Lone I is from Japan, and "Lone" means the never-ending stream of waterfall. Its English name is named after its Chinese homonym "Lone ai", directly expressing the theme of the music bar – "the lonely I" find a temporary habitat here.

Lone I club features an open space layout of "single building near waterfront + space placed outside + waterway across streets", creating a poetic meaning of water flowing beneath a little bridge and man lying down listening to the scull sound under river. The internal business area is more than 1000 square meters, and the main building is divided into two floors, where the first floor is mainly classical European style, open and quiet, with soft and cozy lighting and sound to meet guests' needs of getting together and chatting with others. The second floor is linked to the first one with a built-in elevator, and the independent, trendy and avant-garde modern style is appropriate for business and leisure place.

In the operating model of Lone I, simple meals, slow rock, performing arts and other diversified forms coexist, and the designer finished the creation of the space temperament with the treasonable idea of "resolving conflict with conflict". From materials and lighting to the music setting, "conflict" is raised to the extreme to show a coordinated and natural beauty.

Lone
泷爱
CLUB

EAGLE BAR

泷爱一名，源于日本，“泷”乃川流不息的瀑布之意，后又取名其谐音“lone i”作为英文名，直接地表达了音乐酒吧的主题——“孤独的我”在这里找到暂时的栖息地。

泷爱酒吧以“临水独栋 + 外摆空间 + 穿街水道”的开放式空间组合，营造出小桥流水、枕河听橹的诗情画意。内部营业面积达 1000 多平米，建筑主体分为上下两层，一层以欧式古典风格为主，开放、闲适，灯光音响柔和惬意，满足了客人聚会畅聊的需求；二层与一层以内置电梯相连，独立、前卫、潮流的现代风格用于商务休闲再适合不过。

泷爱的经营方式中简餐、慢摇、演艺等多元化形态共存，设计师以“用冲突解决冲突”的叛逆性思路完成了对空间气质的塑造，从材质、灯光到音乐配置，将“冲突”推向一种极致，呈现出一种协调、自然的美。

ONE
ABUNDANCE
CLUB
SHOUT LIFE
FREE

Pioneer
CDJ-2000
Pioneer

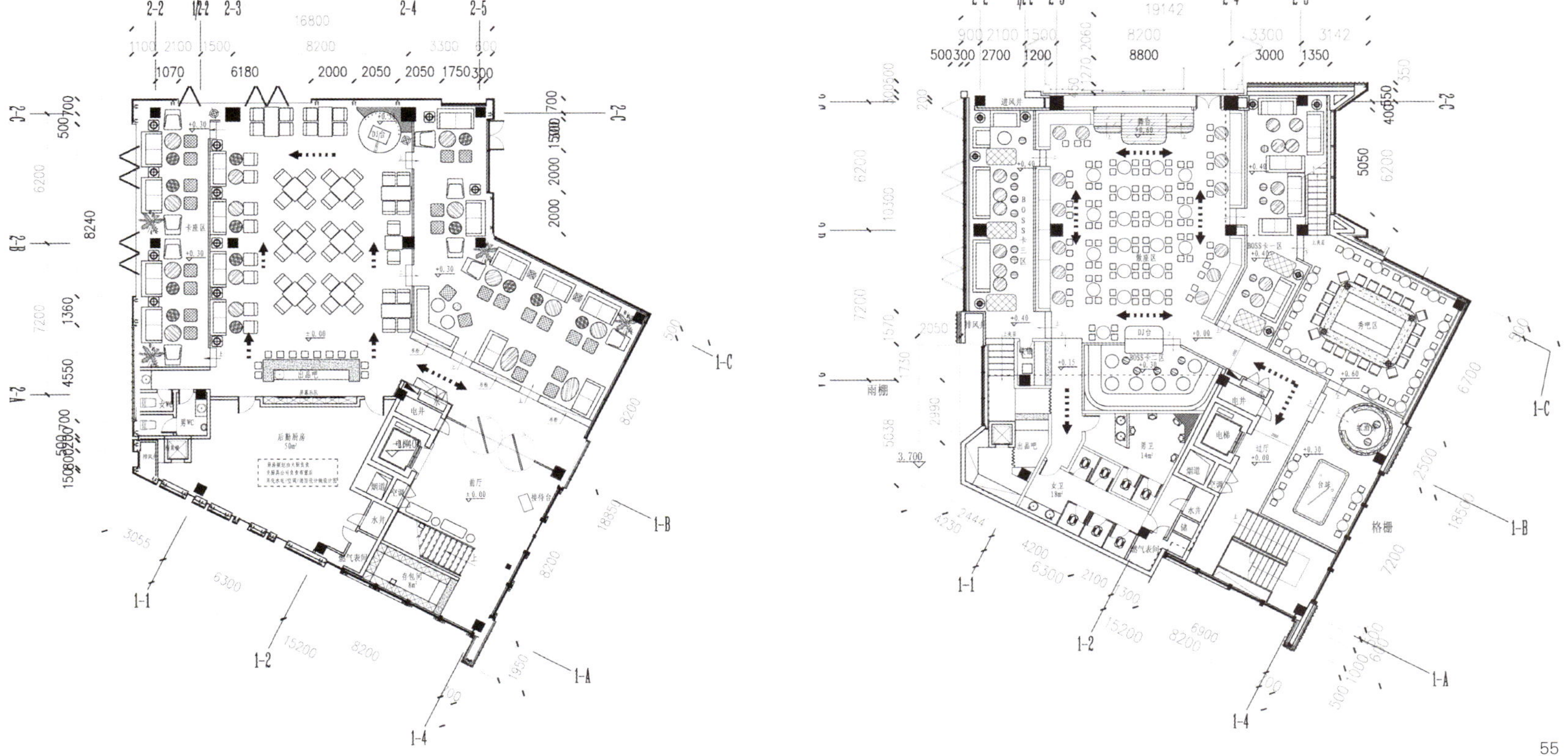

Shenzhen Xiyue Restaurant & Bar (the MixC Store)

项目地址
深圳

面积
700 平方米

设计师
陈武

设计公司
深圳市新冶组设计顾问有限公司

主要用材
墙面波斯海浪灰大理石、地面圣罗兰大理石、旋转大门古法琉璃、浮雕实木搽色做旧、仿古木地板、天花浅香槟金银箔、紫铜、仿古实木板

深圳市喜悦西餐酒吧（万象城店）

The project is located on the third floor of Vientiane City Phase Ⅱ, integrating with restaurant and bar, at the same time, it holds parties, so it features a sense of dining ritual and leisurely sense of bar. The LOGO is not conspicuous, but clear and steady.

The rotating door is blunt and dignified, as if leaving thousands of kinds of worries behind. The leaves made of red copper are scattered on the Persian gray marble with wave patterns, featuring tranquility at first glance. Surprise comes along, a large area of green plants which weave the whole surface into a "breathing wall", accompanied by the gurgling of water, to dispel the sense of alienation from the outdoor concrete and iron forest. A large number of living plants are used in the interior space, which is not only the bold application of new technologies, but also a show of the designer's care and ingenuity. In the impetuous modern city, even a hint of pure green, a breath of natural air, is a solace for people. The tough post-industrial space is filled with vitality, which is also a challenge to the fixed style.

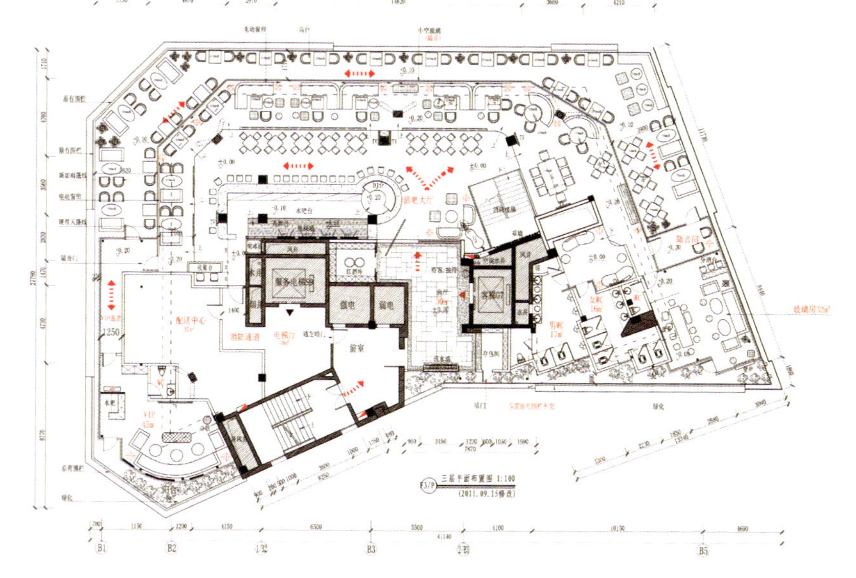

项目位于万象城二期三楼，集餐厅与酒吧两种业态于一体，同时还兼营聚会，因此，兼具用餐的仪式感与酒吧的闲适感。LOGO虽不张扬，却清朗笃定。

旋转大门钝重端庄，仿佛能隔开身后的万千烦忧。紫铜造型的树叶散落在波斯海浪灰大理石上，给顾客第一眼的安宁。惊喜随后而至，大片绿色植被织就成一整面“会呼吸的墙”，伴随着潺潺流水声，消解了室外钢筋水泥森林的疏离感。将活体植物大量运用于室内空间，既是新技术的大胆应用，又传递出设计师的关怀与巧思：在浮躁的现代都市里，一丝纯绿、一点自然生息的空气，都是慰藉。在后工业气质的硬朗空间里注入勃勃生机，也是对模式化风格的挑战。

Guangzhou Vi-good Fashion Restaurant

项目地址
广州

设计公司
深圳市新冶组设计顾问有限公司

设计师
陈武

主要用材
黄橡木防火板、黄橡木饰面板、茶镜、黄色烤漆玻璃、绿色烤漆玻璃、仿木地板、人造石、肌理漆

广州味可餐厅

Vi-good Fashion Restaurant is located in the No.5 parking apron shopping mall of the primary Baiyun Airport in Guangzhou. Here features a strong commercial atmosphere with heavy sense of alienation, therefore, the entrance is designed with a large number of wooden veneers with smooth modeling, where the bright and pristine texture of materials soothes the cold feeling in the airport space, and the appropriate style endows the space with more affinity. The fashionable metal doors are flexible and convenient, not only avoiding the old-fashioned and closed feeling, but also expressing the feeling of open and welcoming the customer, at the same time, it sets aside space for the peak flow.

The indoor space for activities is semi-enclosed, with flexible circulation; the flexible arrangement of natural elements extends the limited space, enhancing the visual sense of space, so the restaurant is transparent, free and flexible. The wall is decorated with plant and stone, coupled with green paint glass, exuding a soft lighting, echoing with the theme. The whole designing technique is neat and simple, allowing guests in the restaurant to forget the toilsome journey and enjoy the delicacies quietly, and let the happy moments become eternal.

Vi-good味可
Fashion Restaurant · 新派餐廳

Vi-good味可
Fashion Restaurant · 新派餐廳

Vi-good味可
Fashion Restaurant · 新派餐厅
2F

Vi-good味可
Fashion Restaurant · 新派餐厅

Vi-good味可
Fashion Restaurant · 新派餐厅

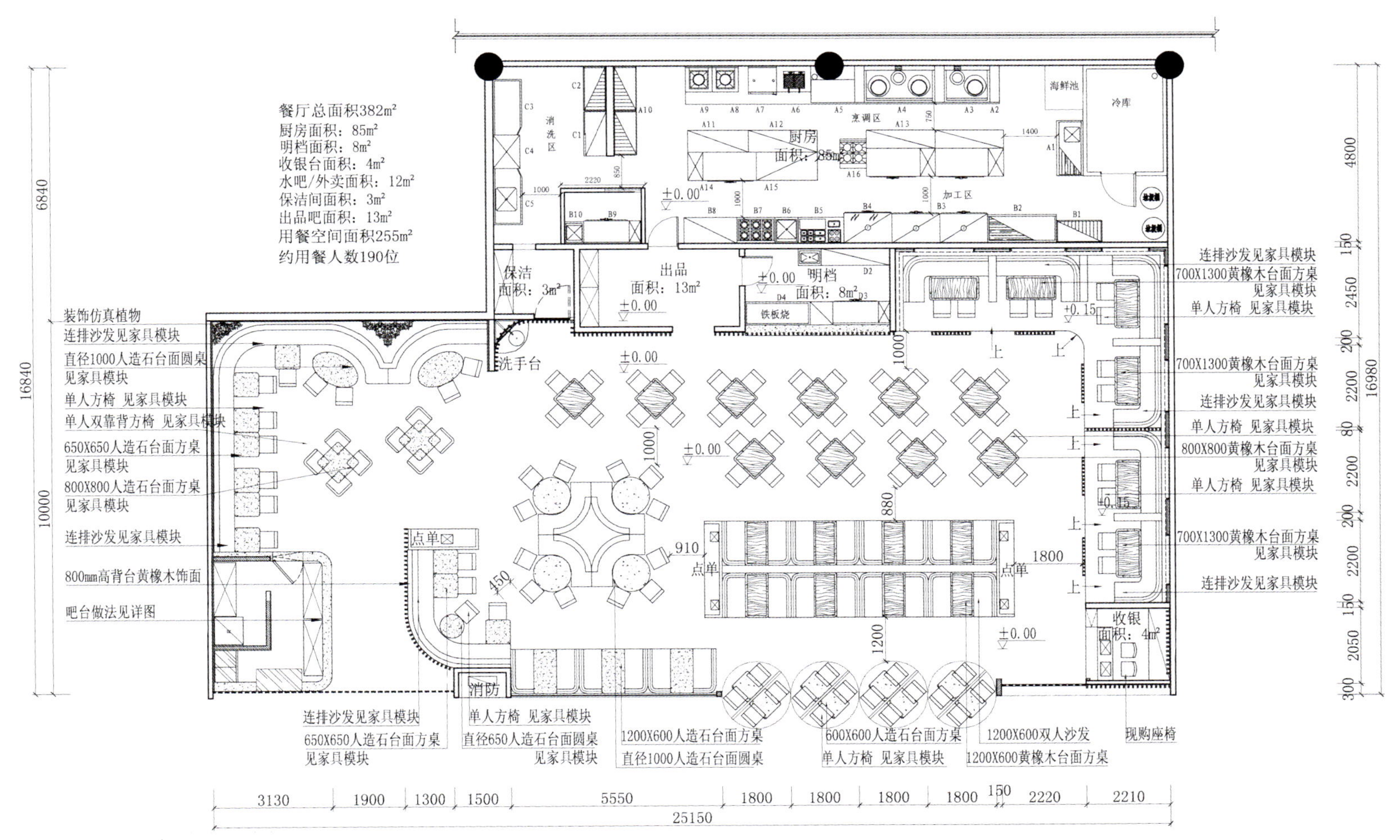
餐厅总面积382m²
厨房面积：85m²
明档面积：8m²
收银台面积：4m²
水吧/外卖面积：12m²
保洁间面积：3m²
出品吧面积：13m²
用餐空间面积255m²
约用餐人数190位
装饰仿真植物
连排沙发见家具模块
直径1000人造石台面圆桌
见家具模块
单人方椅 见家具模块
单人双靠背方椅 见家具模块
650X650人造石台面方桌
见家具模块
800X800人造石台面方桌
见家具模块
连排沙发见家具模块
800mm高背台黄橡木饰面
吧台做法见详图
连排沙发见家具模块
650X650人造石台面方桌
见家具模块
单人方椅 见家具模块
直径650人造石台面圆桌
见家具模块
1200X600人造石台面方桌
直径1000人造石台面圆桌
600X600人造石台面方桌
单人方椅 见家具模块
1200X600双人沙发
1200X600黄橡木台面方桌
现购座椅
连排沙发见家具模块
700X1300黄橡木台面方桌
见家具模块
单人方椅 见家具模块
700X1300黄橡木台面方桌
见家具模块
连排沙发见家具模块
单人方椅 见家具模块
800X800黄橡木台面方桌
见家具模块
单人方椅 见家具模块
700X1300黄橡木台面方桌
见家具模块
连排沙发见家具模块
厨房
面积：85m²
消洗区
烹调区
加工区
海鲜池
冷库
保洁
面积：3m²
出品
面积：13m²
明档
面积：8m²
铁板烧
洗手台
点单
消防
收银
面积：4m²
3130
1900
1300
1500
5550
1800
1800
1800
1800
150
2220
2210
25150
6840
16840
10000
4800
2450
2200
2200
2200
2050
16980

Vi-good味可

味可餐厅位于广州老白云机场 5 号停机坪商场内部。此处商业气息浓厚，疏离感重，因此，入口处大量运用造型流畅的木饰面，以清朗质朴的材料质感舒缓机场空间的冰冷感，妥帖的造型设计令空间更有亲和力。时尚金属门灵活方便，既避免了古板、封闭的感觉，表达出开放、喜迎顾客的心意，又为人流高峰期预留了空间。

室内活动空间做半围合处理，动线灵活;自然元素的自如运用延伸了有限的空间，增加了视觉空间感，令餐厅通透、自由、灵性。墙面采用植物墙和石块元素，加上绿色的烤漆玻璃透出柔和的灯光，呼应了主题。整个设计手法利落、纯粹，让置身于餐厅的宾客忘却了旅途的辛劳，静静地享受美味佳肴，让幸福的瞬间成为永恒。

DANFU LIU
刘 卫 军

PINKI（品伊）创意机构　创始人／执行主席
深圳市品伊设计顾问有限公司　董事总经理
美国 IARI 刘卫军设计师事务所　创意总监、首席设计师

个人特长：
酒店设计、样板房、售楼处、餐饮空间、展示空间、会所设计。以主题式风格创造艺术空间见长，善于融会东西方艺术及文化于生活之中。设计包括：室内设计、陈设艺术设计、家具设计及园林建筑。

社会职务：
CIID 中国建筑学会室内设计分会全国理事
中国建筑学会室内设计分会深圳（第三）专业委员会常务副会长
深圳室内设计协会常务理事
广东省梅州市五华县横陂镇人民政府文化教育发展顾问
首批国家高级室内建筑师
中国百名优秀室内建筑师
中国十佳住宅设计师
中国室内设计师风云人物
全国首批设计行业优秀人才模范
全国设计行业首席专家
中国设计行业特高级研究员
美国 IDCHINA2007 年度封面人物
美国 IDCHINA 室内设计名人堂成员
深圳最具影响力十大室内设计师
“城市荣誉”杰出室内设计师
狮子会荣誉会员
十大高端设计师

Chengdu Ruitaihua Shihao·Jiabai Sales Center

项目地址
成都

面积
594.57 平方米

设计师
刘卫军

设计公司
PINKI 品伊创意机构 & 美国 IARI 刘卫军设计师事务所

主要用材
现代木纹大理石、布朗啡、巴洛克金、球纹桃花芯木饰面、皮革、玫瑰金不锈钢、灰镜、茶镜、墙纸

成都瑞泰华世豪·嘉柏售楼中心

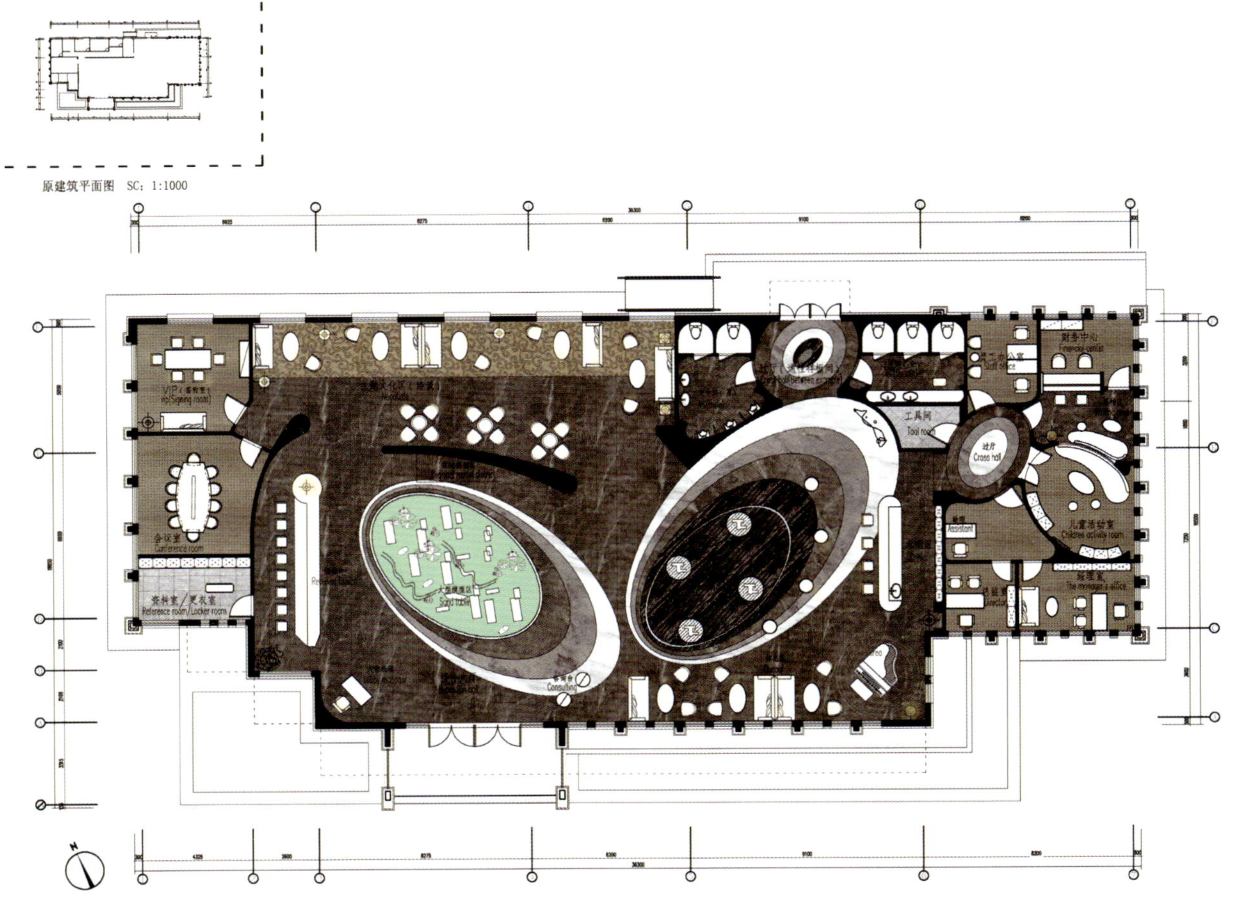

原建筑平面图 SC: 1:1000

The project takes "Urban Charm" as the designing theme, on the base of elegant, stylish and modern designing concept, to comfort the soul and satisfy the needs and aspirations of the urbanites, showing their identity and status.

The design not only integrates modern and fashionable elements, but also interprets the essence of metropolis. The large multi-layered overlapping oval structure on the top of the space echoes with the large one on the ground. The elegant and gorgeous furniture, lighting and crystal clear vases...bright flowers flow in the architectural space of Western charm, in a joyful "bloom", making people in the space appear to be more graceful.

本案以“都市魅影”为设计主题，以高贵、时尚、现代的设计理念为依据，来慰藉都市人心灵的需求和渴望，展现其身份和地位。

设计不仅融入了现代、时尚的元素，更诠释了大都会的精髓。空间顶部的多层重叠大椭圆形结构，与地面多层重叠的大椭圆形形成呼应。典雅、华丽的家具、灯饰和晶莹剔透的花瓶……鲜艳的花丛流淌在西方韵味的建筑空间中，欢乐地“绽放”，令置身其中的人们更显高雅多姿。

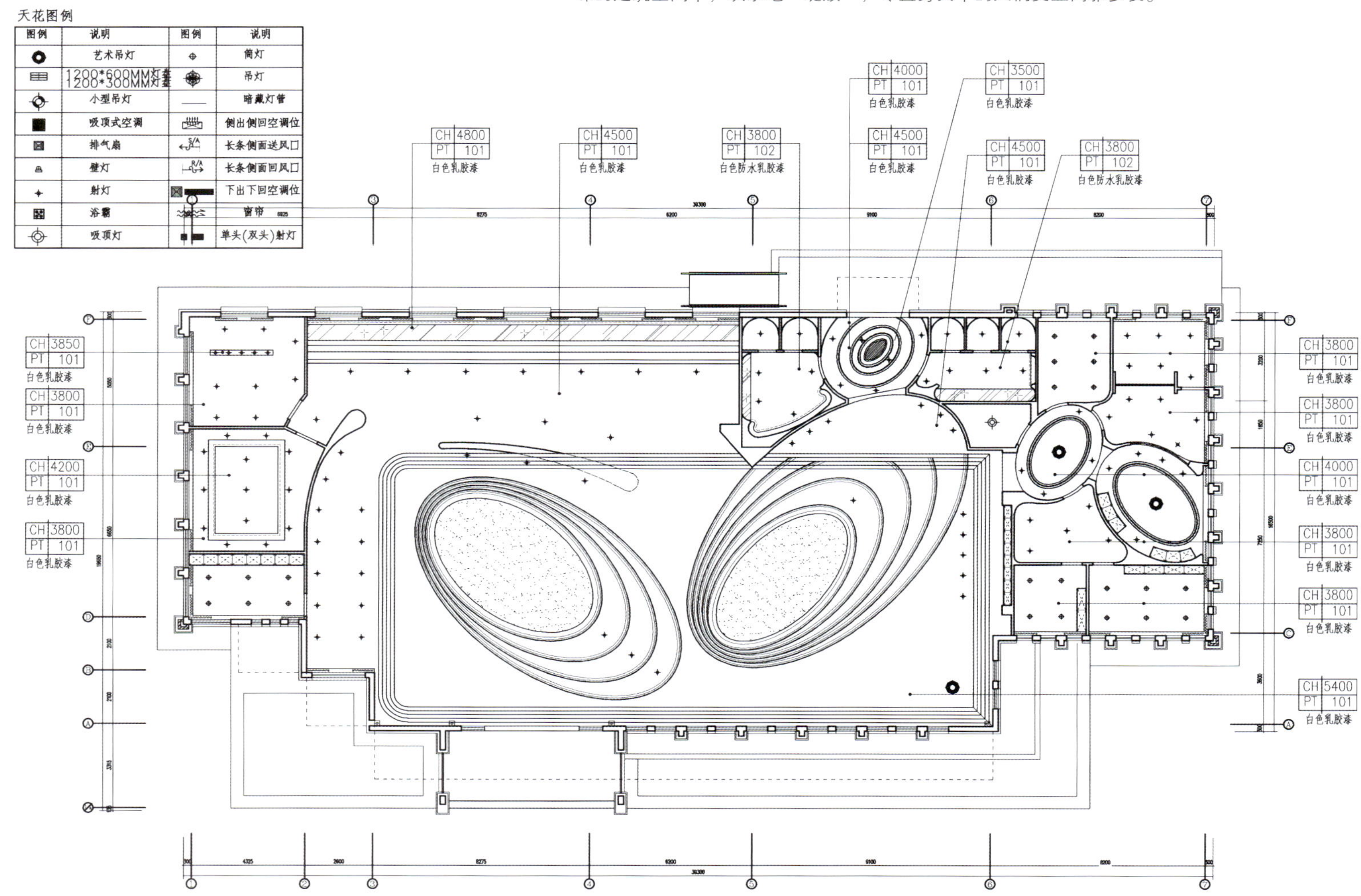

Qinhuangdao Glass Museum Star-grade Fashion Arts Center

项目地址
秦皇岛

设计师
刘卫军

面积
1500 平方米

设计公司
PINKI 品伊创意机构 & 美国 IARI
刘卫军设计师事务所

秦皇岛玻璃博物馆星级时尚文化殿堂

Through the three-dimensional building of the space form, the project realizes the changes of level and depth, laying a solemn and beautiful tone. The tall dome, arch of large span, calm vertical lines, stable and strong pilasters…all of them create an elegant and classical atmosphere, allowing visitors to enter into a European space with profound cultural deposit. Therefore, the design introduces different performing practices according to different forms, including the essences of the Mediterranean, new-classical, Baroque and other styles, coupled with clever space shaping and creative furnishings to realize a new European interpretation to meet the needs of the project.

The business orientation with an integration of various functions, coffee, dining hall and barbecue determines the focus of the project emphasis on realizing the coexistence of rational beauty and sensual beauty, comfortable and luxurious without losing solemn and generous feeling, meeting the multiple needs of high-end business and leisure.

Modern Neo-classical furniture and accessories focus on the contrast and coordination of proportion, color and form, to create an atmosphere intertwined with nobility and elegance, vitality and passion.

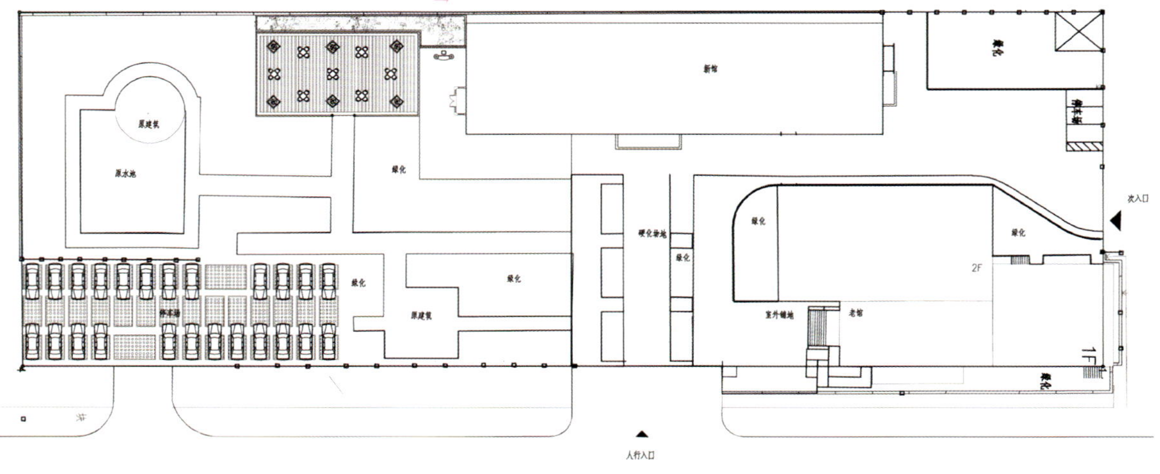

新馆
原建筑
原水池
绿化
停车场
原建筑
硬化场地
次入口
2F
室外铺地
老馆
1F
人行入口

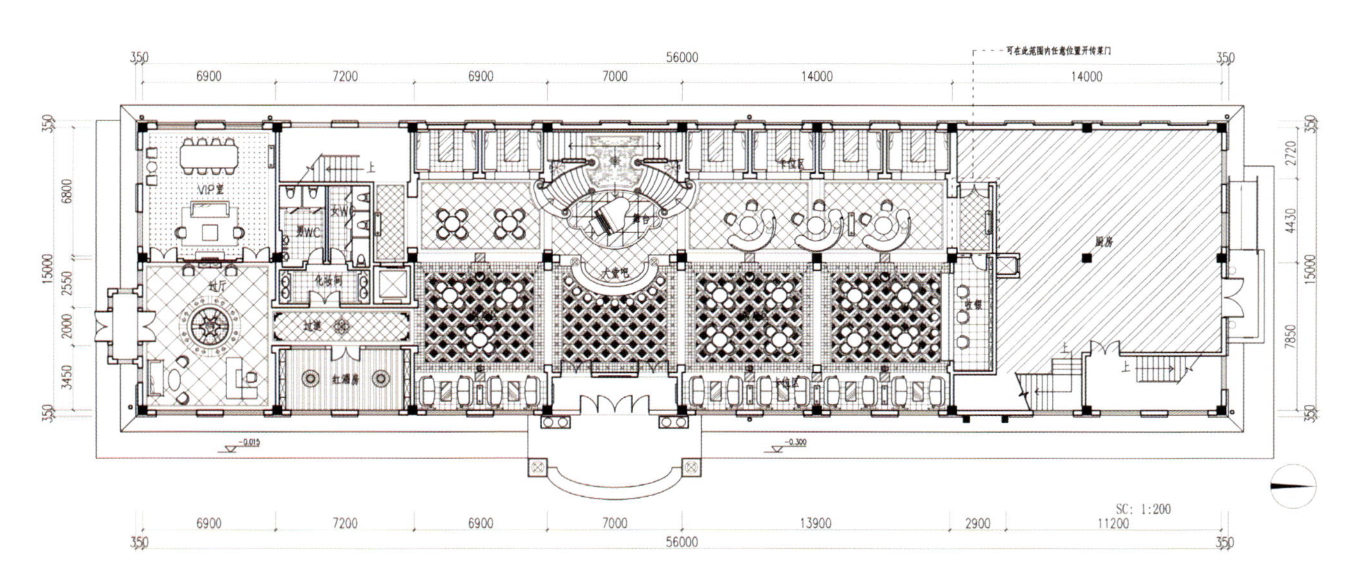

可在此范围内任意位置开传菜门
350
56000
6900
7200
6900
7000
14000
14000
VIP室
上
男WC
女WC
卡位区
舞台
大堂吧
过厅
化妆间
过道
红酒房
收银
厨房
6800
15000
2550
2000
3450
2720
4430
7850
-0.015
-0.300
SC: 1:200
13900
2900
11200

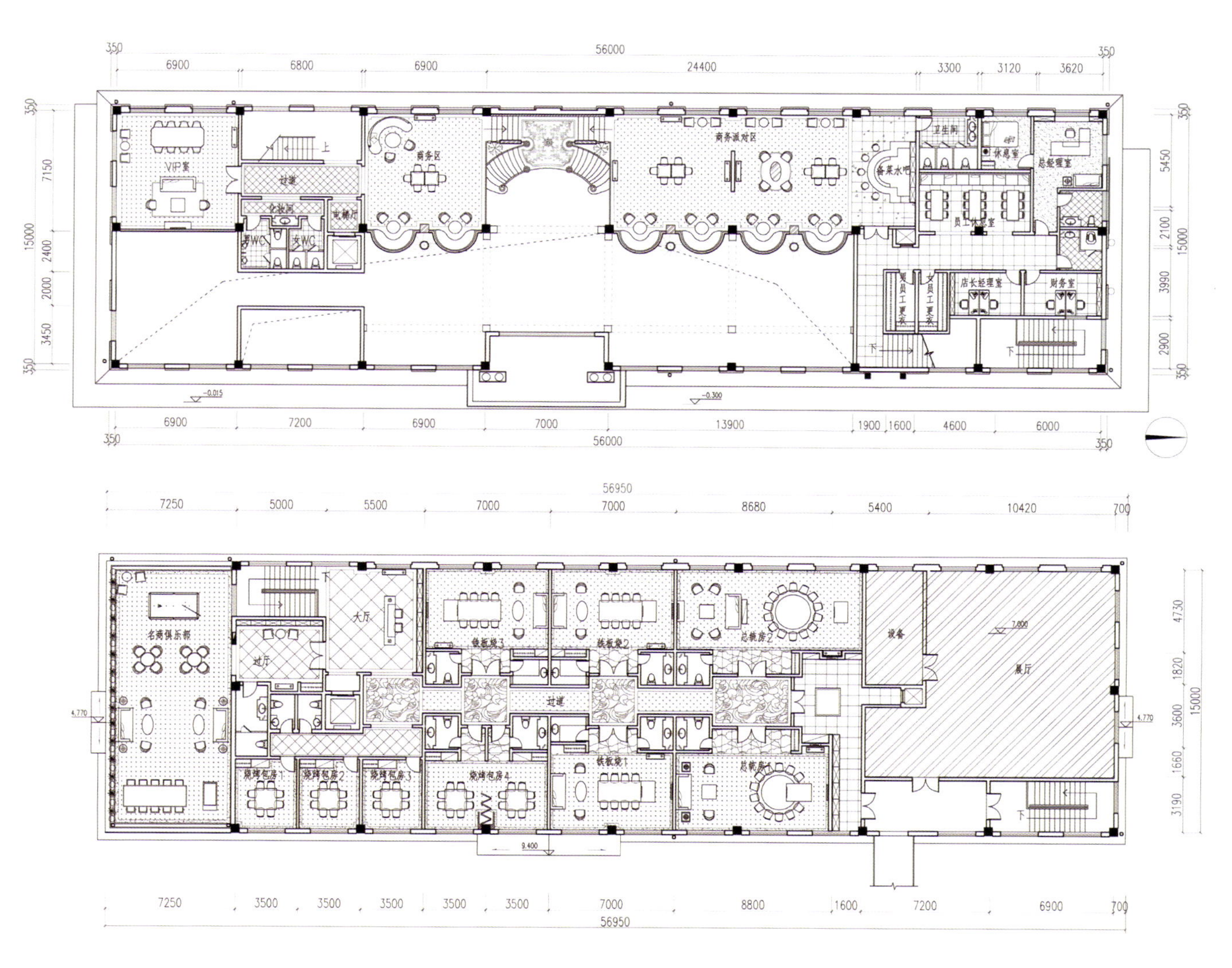

56000
6900
6800
6900
24400
3300
3120
3620
350
VIP室
上
过道
化妆间
电梯厅
男WC
女WC
商务区
商务派对区
备菜水吧
卫生间
休息室
总经理室
员工休息室
男员工更衣
女员工更衣
店长经理室
财务室
下
-0.015
-0.300
7200
7000
13900
1900
1600
4600
6000
15000
7150
2400
2000
3450
5450
2100
3990
2900
56950
7250
5000
5500
8680
5400
10420
700
大厅
过厅
铁板烧3
铁板烧2
铁板烧1
过道
烧烤包房1
烧烤包房2
烧烤包房3
烧烤包房4
总统房2
总统房1
设备
展厅
7.000
4.770
9.400
3500
8800
6900
4730
1820
3600
1660
3190

本案通过对空间形态的立体塑造，实现层次与深度的变化，奠定庄严华美的基调。高挑的穹顶，大跨度的拱券，平静的垂直线条，稳壮的壁柱……营造出优雅、古典的氛围，让人深陷欧洲厚重的文化氛围中。因此，设计根据形态的不同，采用不同的表现手法，其中包括地中海、新古典、巴洛克等风格的精华，再通过巧妙的空间塑造和具有创造性的陈设搭配，来实现符合本案所需的全新的欧式演绎。

咖啡、餐厅、烧烤等多种业态相融合的商业定位决定了本案设计的重点在于实现理性美与感官美的和谐共存，在舒适奢华中不乏庄严大气，满足了高端商务与休闲的多重需求。

现代感的新古典家具及饰品的运用注重比例、色彩、形式的对比协调，营造出高贵与典雅、活力与激情交织的氛围。

Wanfu Royal Palace-Secret Garden

项目地址
浙江平湖

面积
3500 平方米

设计师
刘卫军

设计公司
PINKI 品伊创意机构 & 美国 IARI
刘卫军设计师事务所

万孚尊园——园中秘境

Looking up at the starlit sky, pursuing the rules of our heart, the charm of European culture lies in the yearning for the unknown and the persistence of self-exploration, and art and philosophy, painting and sculpture, music and books comfort and lead people to close to the great and to go towards wisdom.
Inspired by the above concept, this project is designed with the theme of "secret garden", following the trails leading to a secret garden in dream, it presents all one can see and hear in dream with designing language to restore a beautiful illusion. The designer integrates Chinese lifestyle and introduces noble temperament and essence of art, with rich and perfect functions and the elegance and tender as a bright spot, bringing customers a lively space sequence through the smart layout of circulations.

The use of wood, stone and other natural materials fills the air with noble and comfortable atmosphere, so customers feel like wandering in the Louvre, listening to the inner voice of Picasso and Van Gogh, or strolling in the woods of Vienna, or beginning an adventure of Captain Grant...

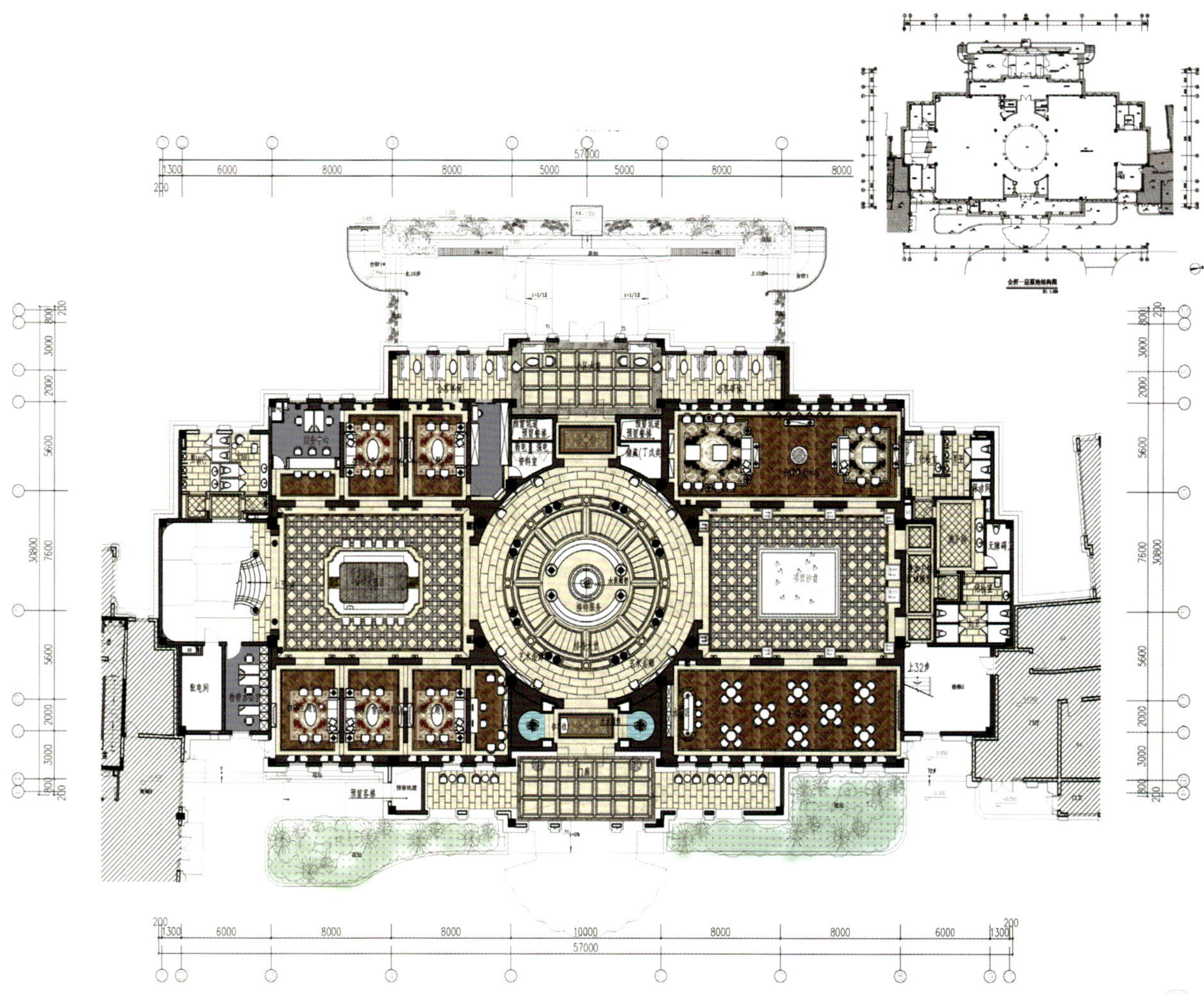

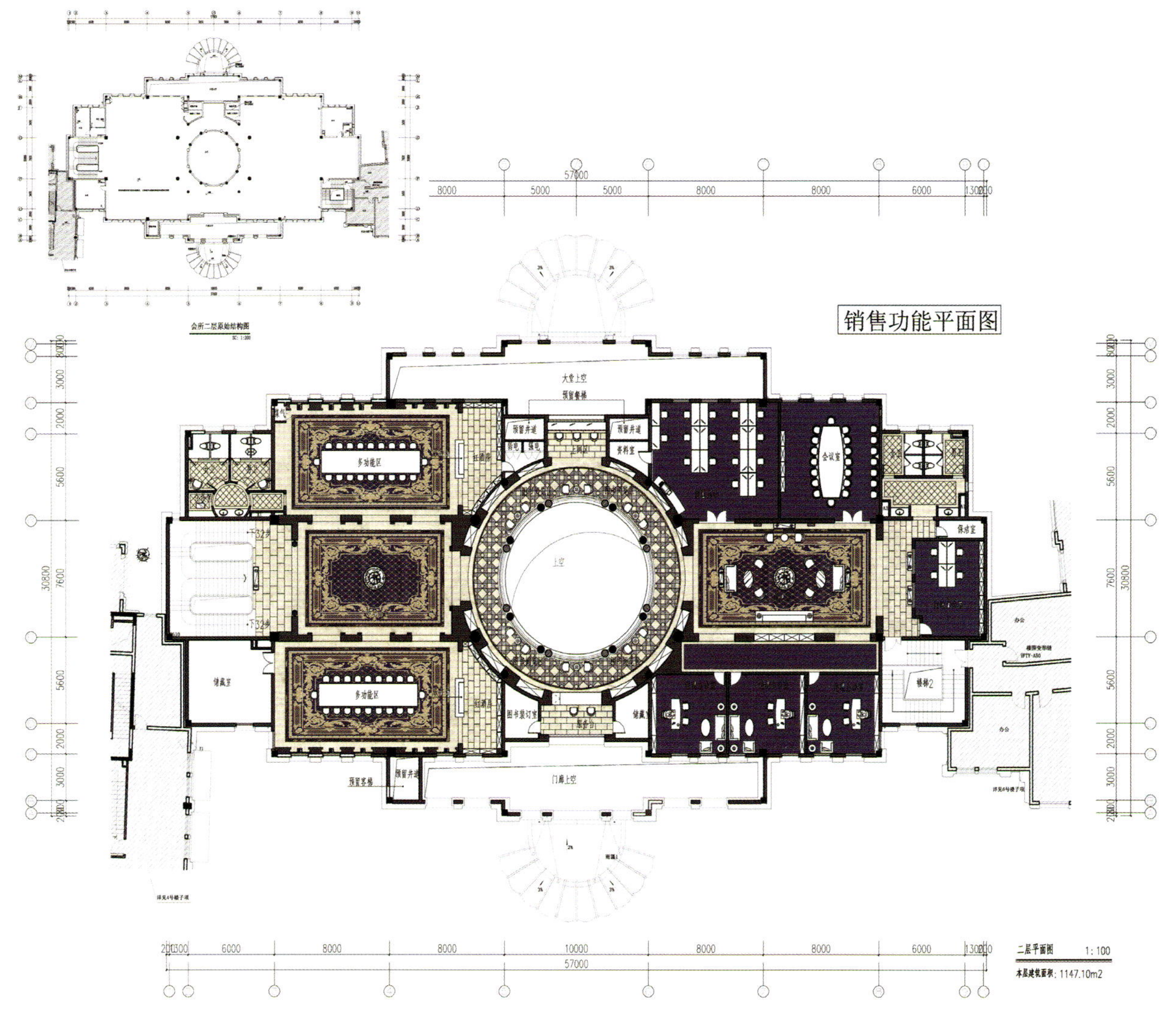
销售功能平面图
会议室
保洁室
楼梯2
办公
二层平面图 1:100
本层建筑面积：1147.10m2

仰望灿烂的星空，追寻内心的法则。欧洲文化的迷人之处在于人们对未知的向往与自我探索的执着，艺术与哲学、绘画与雕塑、音乐与书籍，无不慰藉和引领人类向往伟大、走向睿智。

本案受此启发，以“园中秘境”为设计主题，循着梦境中通往秘密花园的小径，将所见所闻用设计语言呈现出来，还原一个美丽的幻境。设计师融合国人的生活形态、嵌入贵族气质和艺术精髓，将丰富完备的功能和优雅柔和作为亮点，以灵动的动线布局带来活泼的空间序列。木与石材等天然材质的运用令空气中弥漫着高贵与舒适的气息，好像徜徉在罗浮宫中聆听毕加索、梵高的心声，又似在维也纳的森林里漫步，抑或开始了格兰特船长的冒险之旅……

Ziwei Yonghe Square Club

项目地址
西安

面积
2000 平方米

设计师
刘卫军

设计公司
PINKI 品伊创意机构 & 美国 IARI 刘卫军设计师事务所

紫薇永和坊会所

The project takes "integration" as the designing theme, intended to create an inclusive realm, integrating culture with art, producing collision of the classic and modern, and perfectly interpreting the space mood and designing concept.

"Integration" is divided into two floors, where the first floor features model display area, negotiation area, water bar and others, and the designer takes use of the architectural pillars in the original places to decorate each space with a generous and symmetrical axis relationship, dividing the space into different functional areas to enhance the convection and transition between spaces, which is practical and beautiful, showing the concept of "integration." The ceiling of the entrance lobby is designed with the performing technique and texture of mountains and water, integrated with the droplet-like bubble lights to form picture of droplets in the air, at the same time, the bubble lights are applied to every corner of the space to light up the whole space. The second floor features VIP room, office, finance room and conference room, neatly arranged and coupled with wooden veneers, carpet of plain pattern and simple furniture, further highlighting the purity of the space and the integration and collision of various cultures.

本案以“融”为设计主题，意在营造一种包容的境界，将文化与艺术相结合，让古典和现代相碰撞，完美地诠释空间意境与设计理念。

“融”分两层，一层为模型展示区、洽谈区、水吧休闲区等，设计师利用原建筑柱子的网点，使各个空间形成大气而又对称的轴线关系，将空间分割成不同的功能区域，从而增强各空间之间的对流及过渡，既实用、美观，又体现了“融”的理念。入口大堂的天花采用山与水的表现方法和纹理，与水滴般的气泡灯融合在一起，构成空中滴水的意境，同时，气泡灯又延伸至空间的各个角落，点亮了整个空间。二层为VIP室、办公室、财务室、会议室等，整齐的排列，配以木饰面、素纹地毯和简约的家具，更突显出空间的纯净和各种文化的融合、碰撞。

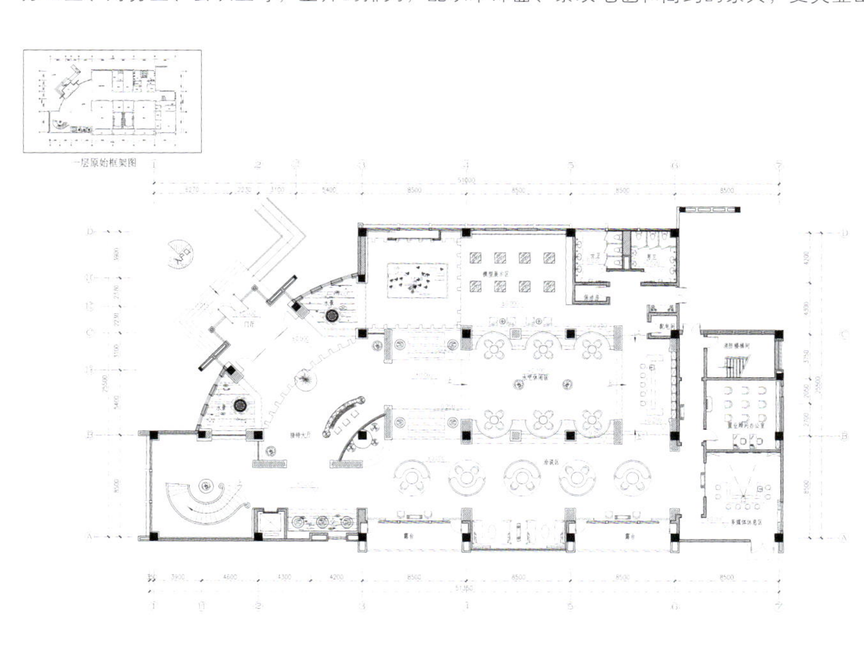

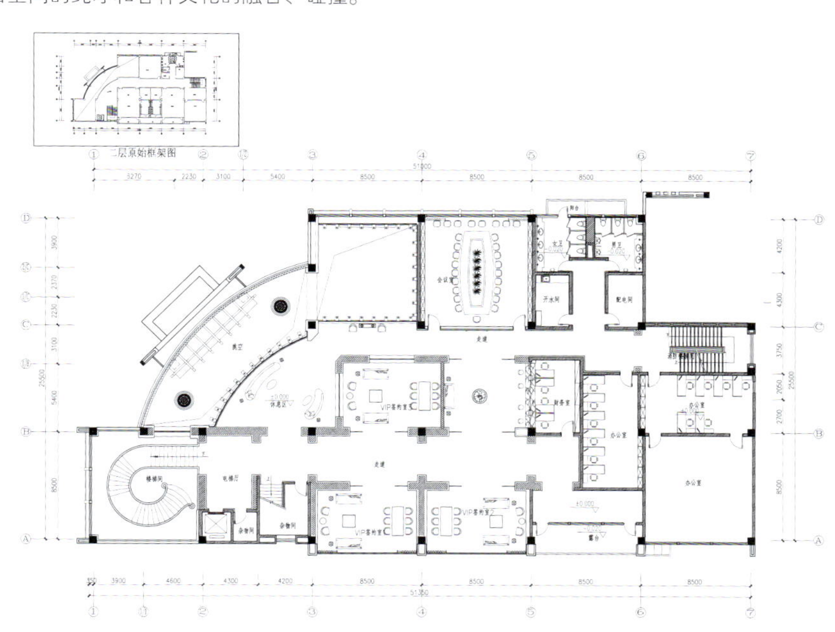

RAYNON CHIU
邱 春 瑞

深圳大易室内设计有限公司　创意设计总监
籍贯：中国台湾高雄，主张“用建筑的角度来做室内设计”
深圳市室内设计师协会理事
国际室内建筑师／设计师联盟理事
国际室内装饰协会理事会员

工作经历：
1986 年　台湾东方设计学院　建筑设计系
1991 年　台湾大木设计师事务所设计师
1994 年　美国李奥国际设计事业主任设计师
1999 年　台湾亚太设计驻中国大陆设计总监
2005 年　成立深圳大易室内设计有限公司
2010 年　成立台湾大易国际设计事业（香港）有限公司

代表作品：
中信惠州汤泉龙泉居温泉度假会所
中信集团深圳红树湾空中顶层会所
北海天隆三千海高尔夫会所酒店
海口时尚桌球运动会所
新疆昌吉农科展示中心
西安时代酒店
四川德阳大酒店
邵阳宝庆山庄大酒店

Beihai Tianlong Three Thousand Sea Golf Club

项目地址
广西北海

面积
9600 平方米

设计师
邱春瑞

设计公司
深圳大易室内设计有限公司

主要用材
海浪灰、意大利木纹、古木纹、鱼肚白、黑伦金、雨林绿、亚洲米黄、金香玉、黑金砂、深啡网、铁艺、银镜、马赛克、木饰面板、实木地板、塑木地板

北海天隆三千海高尔夫会所

The exterior wall of the club is splendid and spectacular, and the interior decoration is graceful and luxurious, generous and exquisite, featuring post modern European style. The club features large dressing rooms, golf specialty shop, Chinese and Western restaurants, saunas, hot spring SPA guestrooms, outdoor hot spring pools, leisure tea bar, cigar bar, red wine bar, providing banquet reception and other services.

The Chinese and Western restaurants, Coffee bar, cigar bar and red wine bar are located on the first floor, which is decorated with ancient and refreshing style. The dressing rooms are on the right side, divided into male and female parts, featuring wet and dry steam rooms and large hot spring pools. The second floor is designed with a large 300-seat banquet hall, 14 large or medium-sized private rooms, which can undertake large banquet, BBQ and cocktail parties. In addition, it features a leisure tea bar with balcony of floor-to-ceiling windows for a panoramic view. Guests and friends fill up the club, enjoying a cup of tea (named Jiuqu red plum), bathing in the sunshine in stadium, allowing sea breeze to blow over long hair, such experience is so beautiful and romantic.

The guestrooms are placed on the third floor with full facilities and quiet environment. Each room is designed with large hot spring pool which is seldom seen in other places, so this place is the first choice for golfers to enjoy a luxurious experience.

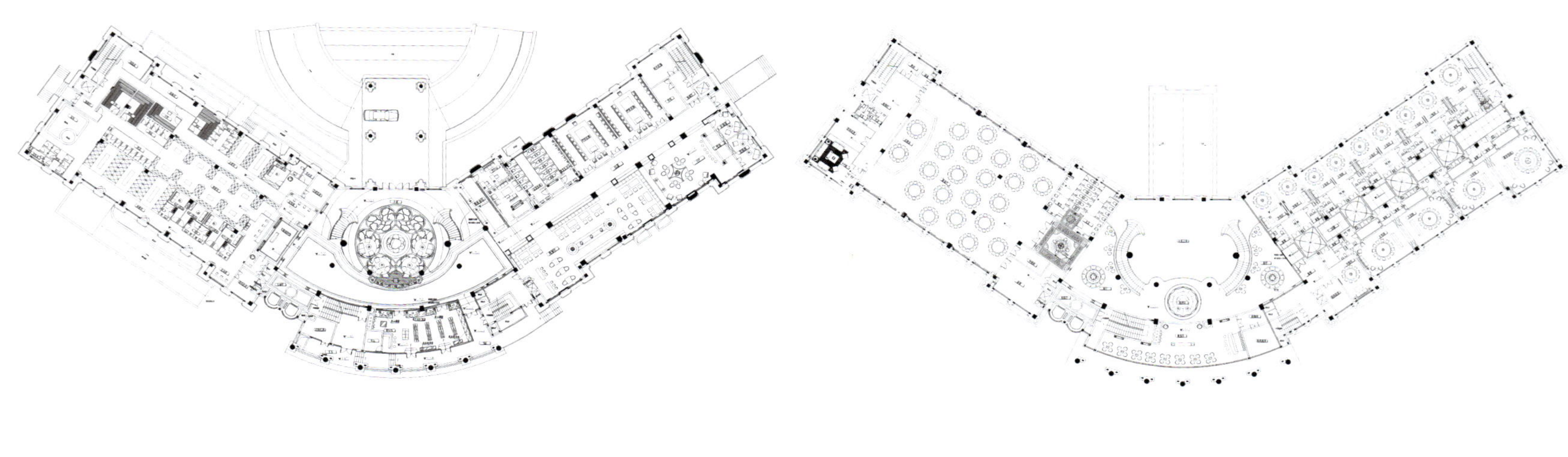

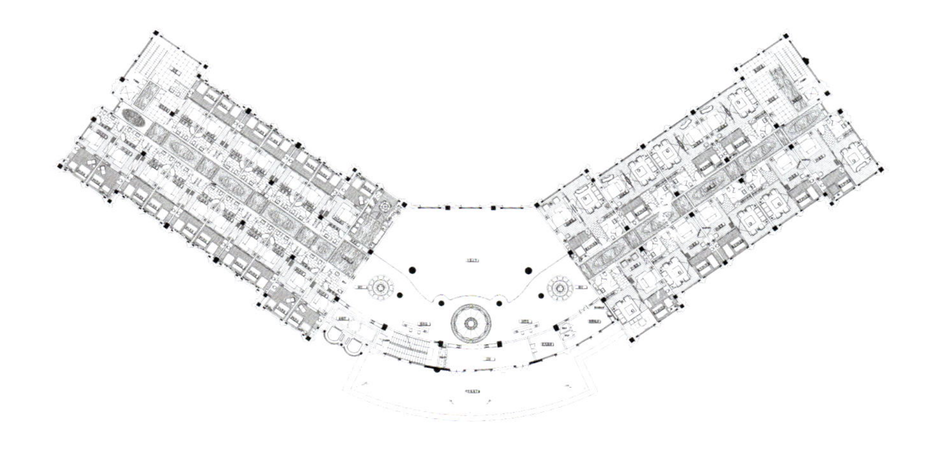

会所外墙气派壮观，内部装修雍容华贵、大气精致，呈后现代欧式风格。

会所配有大型更衣室、高尔夫专卖店、中西餐饮、桑拿、温泉 SPA 客房、室外温泉泡池、休闲茶吧、雪茄吧、红酒吧，提供宴会接待等服务。

中西餐厅、咖啡厅、雪茄吧、红酒吧分布在一楼，装修古朴而清新。更衣室位于右侧，男女分区，设有干、湿蒸房和大型温泉泡池。二楼设有一个可容纳三百人的大型宴会厅、14 个大中包厢，可以承接大型宴会、BBQ、鸡尾酒会。此外，还有一个落地窗阳台式全景休闲茶吧，高朋满座时，品一杯九曲红梅，沐浴球场阳光，任海风拂动长发，唯美而浪漫。

客房设于三楼，配套齐全，环境清幽。每个房间都设有极少见的大型温泉泡池，是球友入住和奢华体验的首要选择。

宴会厅

Zhangzhou Junyue Gold Coast Show Flat Unit B1

项目地址
漳州

面积
93 平方米

设计师
邱春瑞

设计公司
深圳大易室内设计有限公司

主要用材
爵士白、黑白根、西班牙米黄、意大利木纹、银镜、木地板

漳州君悦黄金海岸样板示范单位 B1 户型

Hundreds of thousands of people are obsessed with the snowy castle in the Nordic fairy tale; the pure and airy beauty can put peace and calm deep into your soul. The designer turns such kind of beautiful image into reality.
The holy white becomes the protagonist of the space, coupled with the irradiation of lights and reflection of translucent mirrors to create a spacious and bright living space. The concise European-style design abandons complex traditional European style, with concise and smooth lines cutting out geometries one by one to ease the sense of monotony in the white space, making the whole space more three-dimensional, and endowing the interior space with rich expression. The elegant and exquisite decorations are introduced to show a life attitude which is to constantly pursue high quality of life. A few touch of black comes out occasionally and brings you with noble air, coupled with the white space to show a classical artistic effect.

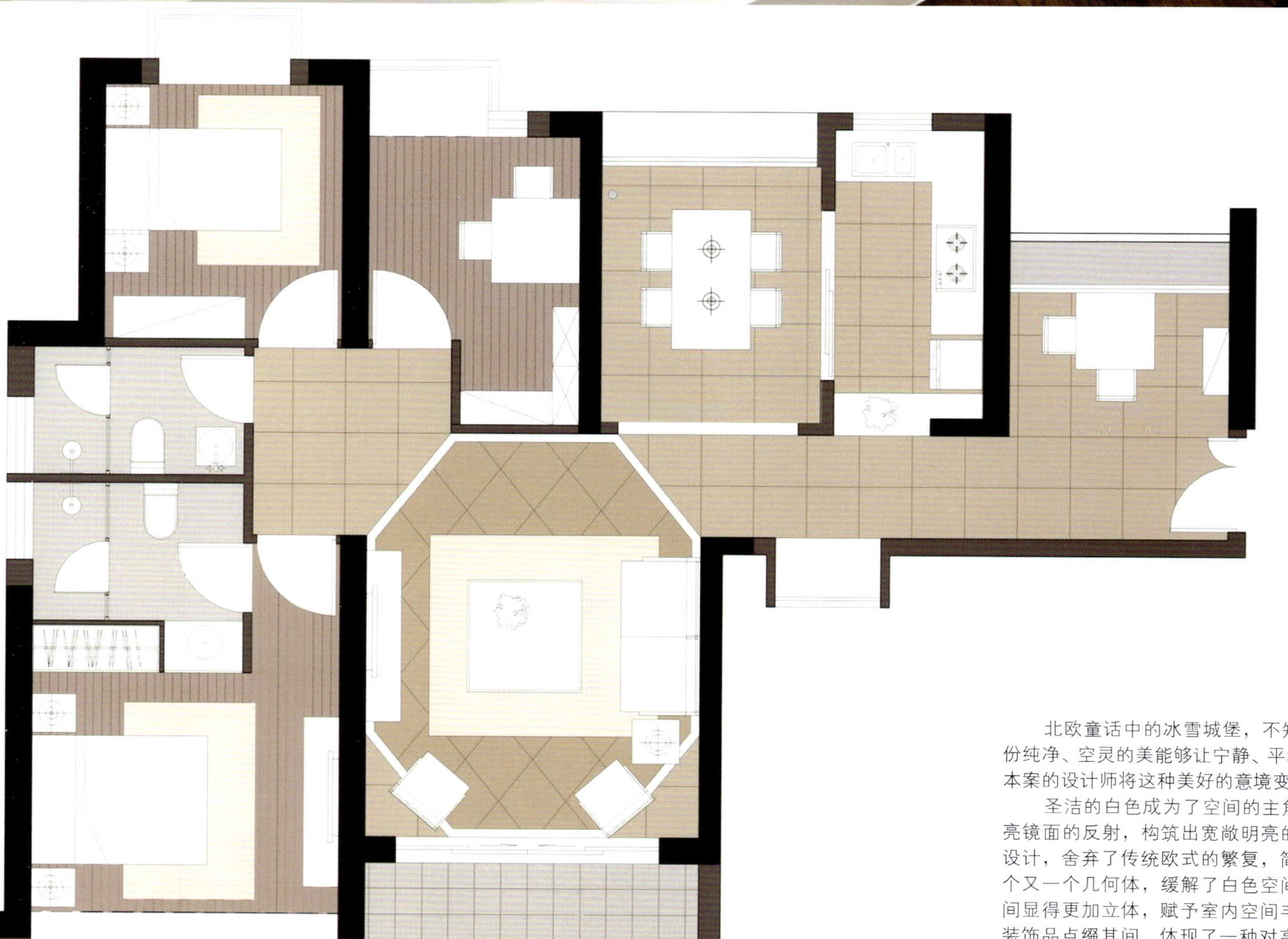

北欧童话中的冰雪城堡，不知俘获了多少人的心，那份纯净、空灵的美能够让宁静、平和深入到人的灵魂深处。本案的设计师将这种美好的意境变成了现实。

圣洁的白色成为了空间的主角，配合灯光的照射、透亮镜面的反射，构筑出宽敞明亮的生活空间。简欧风格的设计，舍弃了传统欧式的繁复，简洁流畅的线条切割出一个又一个几何体，缓解了白色空间的单调感，并让整个空间显得更加立体，赋予室内空间丰富的表情。精美考究的装饰品点缀其间，体现了一种对高品质生活不懈追求的人生态度，偶尔出现的几抹黑色，带来高贵的气息，与白色空间相搭配，展现出经典的艺术效果。

Xi'ning Hengchang Louvre Mansion Sales Center

项目地址
青海西宁

面积
1600 平方米

设计师
邱春瑞

设计公司
深圳大易室内设计有限公司

主要用材
阿富汗黑金花、咖啡金、流金啡、西班牙米黄、金箔、爵士白、黑钛金、铜条、银镜、灰镜、皮革、黑色铁艺

西宁恒昌卢浮公馆营销中心

The construction of the project and the outdoor landscape are defined with French style, which is gorgeous and elegant. Therefore, the designer firmly holds two key words, "luxurious" and "generous", to create this show flat.

The large staircase made of high-grade stone and exquisite wrought iron, carefully carved plaster ceiling, magnificent Roman columns, brilliant large-sized crystal chandeliers, large colorful oil paintings, spacious and open space for use, exquisite floor stone with various patterns...all the elements are combined together to the right, to create a distinguished and generous space frame. Then the designer selects some luxurious and exquisite furnishings to decorate the space, such as gorgeous European-style sofa, elegant grand piano, and chic decorating utensils... influencing everyone in the room by its momentum, and allowing clients to feel the style of the real estate on sale, which caters to the brand positioning of Louvre Mansion.

本案的建筑和室外景观都以法式风格来定义，华丽而优雅。因此，设计师紧紧抓住"奢华"与"大气"这两个关键词来打造这个展示空间。

高档石材与精美铁艺相结合的大型楼梯，精雕细刻的石膏天花，气势磅礴的罗马柱，璀璨夺目的巨型水晶吊灯，色彩绚丽的大幅油画，宽敞辽阔的使用空间，地面上花纹多样的精美石材……恰到好处地组合在了一起，构建出一个尊贵大气的空间框架。然后，设计师选用了一些华贵精致的陈设填充其中，如华丽的欧式沙发，气质优雅的三角架钢琴，别致的装饰器皿……让所有进来这个地方的人都能被它的气势所感染，感受到其所销售的楼盘的格调，迎合了卢浮公馆的品牌定位。

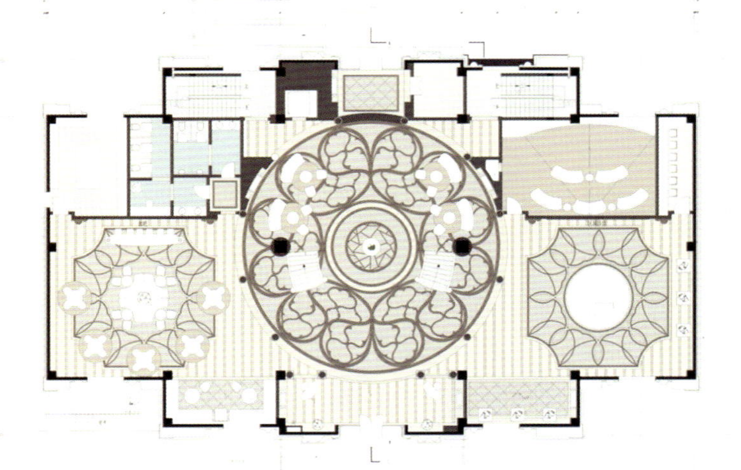

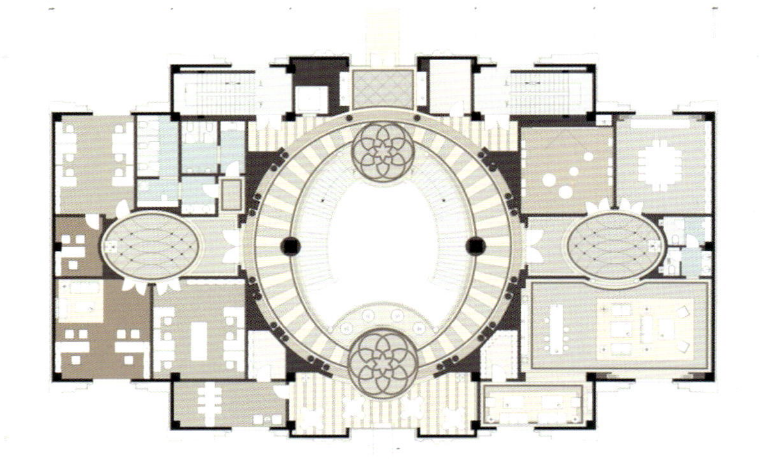

SHUHENG HUANG
黄 书 恒

玄武设计主持人　中国台湾知名建筑师／设计师

黄书恒先生为中国台湾最大建筑商远雄集团的长期合作对象，参与其系列造镇计划。致力于现代科技与传统哲学的完美结合，精于铺排空间、锻造细节，创建风格独特的产品，客户遍及两岸，为高端房产销售的有力推手。曾获日本 JCD 商业空间大赏、现代装饰国际传媒奖（年度样板空间大奖／年度最佳展示空间）、IAI 亚太国际室内设计菁英奖、金堂奖、海峡两岸四地室内设计大赛住宅建筑类／公共空间类大奖等。

最新作品集《In Search of eternity——黄书恒建筑师·玄武设计隽品集》于 2012 年 12 月正式出版发行。

玄武设计网址：http://www.sherwood-inc.com

E2 Xizhi Sales Center

项目地址
中国台湾

设计师
黄书恒

设计公司
玄武设计

E2 汐止接待中心

Constrained by the long and narrow terrain, the main building presents a long and narrow shape with a total length of 200 meters and only a width of 10 meters, just like a huge dragon on the ground waiting to soar. The tortuous stature plays a role of ecological fence to separate it from the blunt and cold surrounding environment. Looking the building from afar, you can see a large exterior wall standing there, and the facade is designed with large-sized glass to create an airy and hazy feeling. In contrast with the green color, the building looks like a mosaic forest, containing the important concept of ecological conservation, and allowing the building to be integrated seamlessly with the surrounding living environment. At night, the wall reflects changing colors, pinkish purple, yellow, blue...which is so colorful and it decorates the vision and mind of each visitor. The sense of technology is the designing axis of space. Simple but not stiff pure white allows thoughts to fly freely, not only maintaining the advanced sense of architecture, but also cleverly ablating people's worries about the "depersonalization" of reinforced materials.

NORTH AMERICA
美國 UNITED STATES
MID AMERICA
Credits,Farglory

遠雄
Farglory
U·TOWN

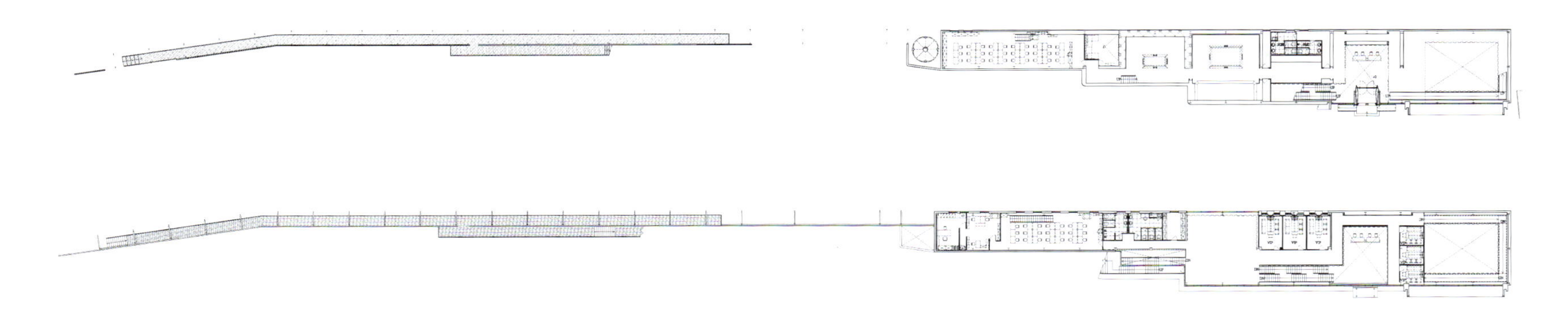

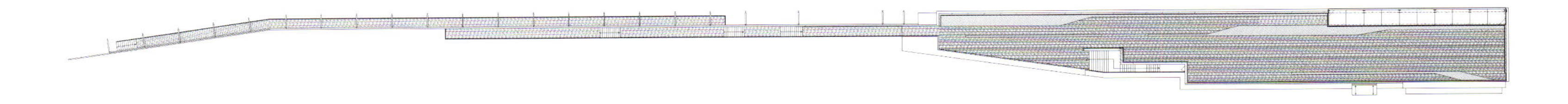

受狭长地势的限制，建筑主体呈总长200米、面积却只有10米的狭长型，犹如一条蛰伏于地面、等待飞升的巨龙，曲折的身形也如同一道生态围墙，隔绝了周遭环境的生硬与冰冷。远远看去，可见大片墙面伫立眼前，立面采用大片玻璃，营造轻盈朦胧的感觉，于青绿色泽的衬托下，建筑犹如被马赛克化的森林，既蕴含着生态保育的重要观念，又让建筑与周围的居住环境无缝接合。夜晚时，墙面可切换不同的色泽，粉紫、黄、蓝……色彩纷呈，点缀着每个访客的视野与心灵。

科技感是空间的设计主轴，简单而不生硬的纯白色让人们的思绪自由飞扬，既可维持建筑的先进感，又巧妙地消除了人们对钢筋材料“去人性化”的忧虑。

Bolon Art Co., Ltd Office

项目地址
中国台湾

设计公司
玄武设计

面积
245 平方米

主要用材
波龙毯、壁纸、玻璃隔断

设计师
黄书恒、许棕宣、陈昭月

摄影
王基守

波龙艺术有限公司办公室

"Bolon Art" is well known for its modern sense and natural style, with its exclusive technology to present a complicated texture, allowing users to hover between the actual and the virtual through the performence of colors and lines.

The wall and ceiling are decorated with white color, infiltrating visitors' senses with bright and light feelings, and the continually convoluting white lines look like white swirl leading visitors to spin into the art of illusion. The carpet of earth tone covers the negotiation space, echoing with the horizontal carpet on the wall to bring visitors a feeling of "down to earth". The main office area is separated from the outside space by blue glass, differentiated it from the large area of white in the entrance hall, and also connected to the gray main office area, maintaining the easy sense of negotiation with visitors, without breaking the rigorous temperament of office area.

BOLON

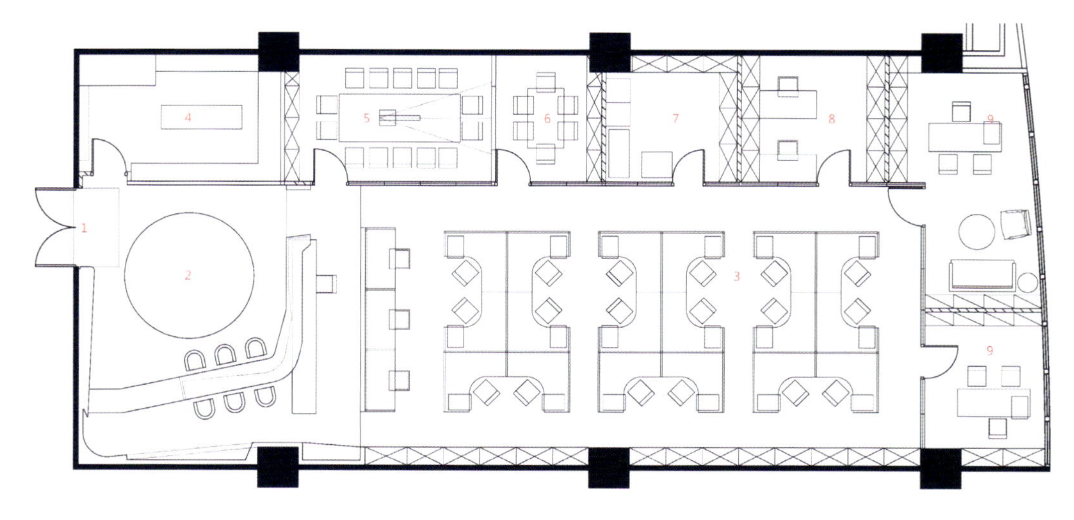

波龙办公室平面配置图

1.入口 2.展示区 3.办公区 4.储藏室 5.大会议室 6.小会议室 7.事务办公室 8.财务室 9.主管办公室

"波龙艺术"以现代感与自然风格闻名业界，利用独家技术呈现繁复纹理，让使用者藉由色彩与线条的轻舞，徘徊于真实与虚幻之间。

设计师用白色装点墙面与天花，以明亮与轻盈感浸润访客的感官，不停旋绕的白色线条犹如白色的涡流，让访客如同旋入艺术幻境。大地色调的织毯铺满洽谈空间，与墙面上的横向地毯相呼应，给访客"脚踏实地"的感觉。主要办公区选用蓝色玻璃分隔内外空间与门厅的大片纯白，亦连接着主要办公区的铁灰，既保持了与访客商谈时的轻松感，又无损工作区应有的严谨气质。

BOLON

U-PARK Sales Center

项目地址
中国台湾

面积
3960 平方米

设计师
黄书恒、欧阳毅、詹皓婷、陈新强、蔡明宪

软装布置
胡春惠、胡春梅

设计公司
玄武设计

主要用材
马赛克、中空板、密底板雷射切割、手工地毯、木作喷漆、铝塑板

摄影
王基守

远雄海德公园接待中心

In Hyde Park building project, Farglory starts U-PARK new town plan to introduce advanced green technology, integrating wind, light, water, green plants and other natural elements with the construction, and the new town of green energy allows the second phase of Farglory house to upgrade again, creating a sustainable building integrated with the environment. In addition to pass on the new vision and new thinking of the building, Farglory Hyde Park Reception Center also undertakes a mission of positioning and updating the corporate image.

Into the shining and colorful box-like building, the V-shaped entrance hall first breaks the visual stereotype of horizon and right angles. The solar panel-like glass mosaic sloping walls on both sides and the staircase made up of steel folded plates extend upward, to create a fun of climbing up. The first floor is designed with a theme hall, where the perfectly round and classical space gradually softens the impact from the architecture appearance. The ceiling is hollowed on eight directions to allow the brilliant daylight to spread from slits, so the white hollowed club enjoys the unique nature, existing and showing the beauty and reality of the building with an overwhelming unique aesthetic.

U-PARK

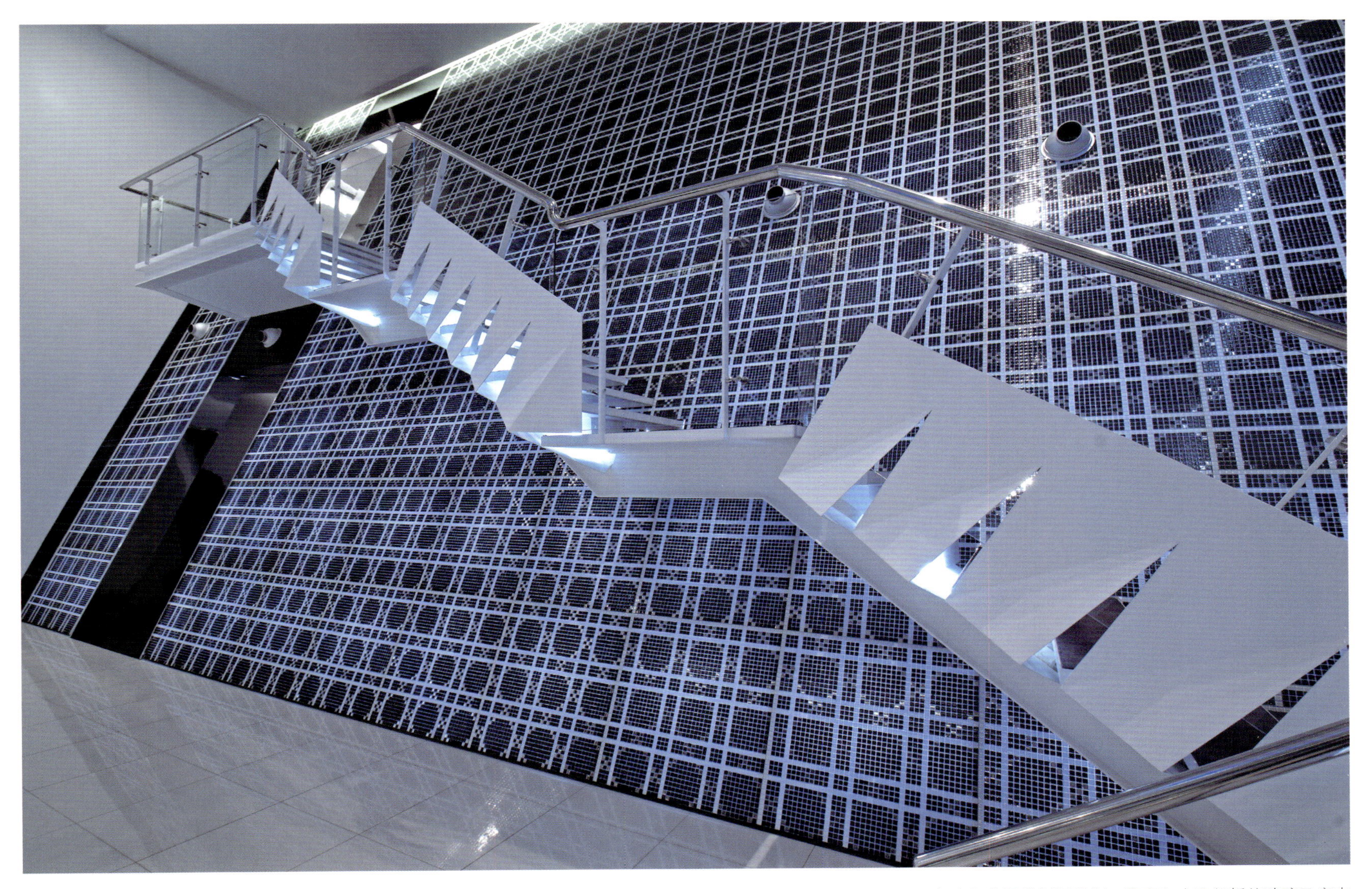

在海德公园建案中，远雄启动U-PARK造镇计划，引进领先的绿色环保科技，将风、光、水、绿等自然元素与建筑结合起来，以绿能造镇，让远雄二代宅再次升级，建构了一个与环境相结合的永续建筑。除了传递建筑新视野与新思维，远雄海德公园接待中心还担任着企业形象定位与更新的使命。

进入光雕彩盒般的建筑中，V形入口大厅首先打破水平与直角的视觉刻板限制。两旁如太阳能板的玻璃马赛克斜墙和钢板折板楼梯向上延伸，营造出倾身攀岩的登高情趣。一楼另有主题馆，浑圆古典的空间逐渐柔和了建筑外观带来的冲击。天花板八方天幕的挑空造型，在缝隙中洒落绚丽天光，让白色镂空装饰的会馆独享自然风华，以压倒性的独特美学存在诉说建筑的美与真实。

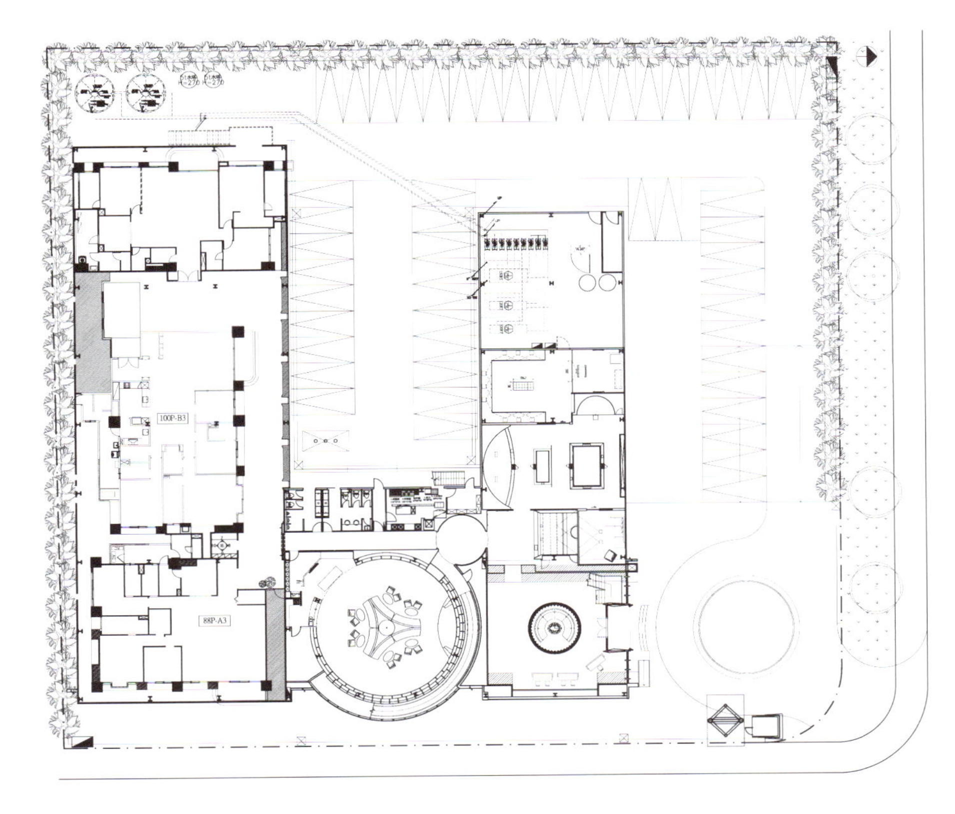

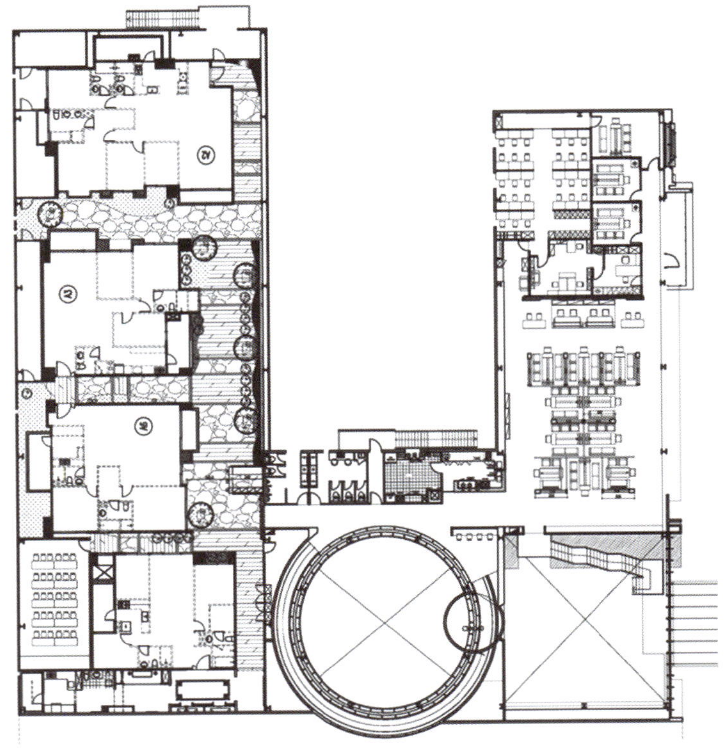

Farglory Marketing Sales Center in Shanghai

项目地址
上海

设计公司
玄武设计

面积
148.5 平方米

主要用材
镜面不锈钢、钢琴烤漆、墨镜喷砂

设计师
黄书恒、欧阳毅、陈新强

摄影
王基守

远雄上海行销展示中心

The project is a touchstone for Farglory to enter into the mainland market. The design takes the steel armor of transformers as the main element of the space, integrating cutting-edge interactive devices, together with the co-construction of hardware and software to practice the dream of building a beautiful and high-tech house.

At the entrance, inspired by the precious acting principle of mechanical watches, the designers design an interactive device called "pupil of future", with the help of camera aperture to start the device when visitors go through the entrance, then the pupil-like device in the middle will open, with the special sound and light effects to lead visitors to explore the convenient way of life in the future.

The interior exhibition space is mainly blue on behalf of the corporate image, and the corporate logo is designed at the entrance, surrounded by three-dimensional templates of armor shapes to demonstrate the concept. The red movable round chairs enclose out a S-shaped seating area, leading the moving lines of the space. While considering the space function, the designers introduce aesthetic style, where rich interactive visual elements are used to create a flexible dialogue between space and man.

FARG

品牌價值永恆
21世紀，全球4大挑戰
環境共生

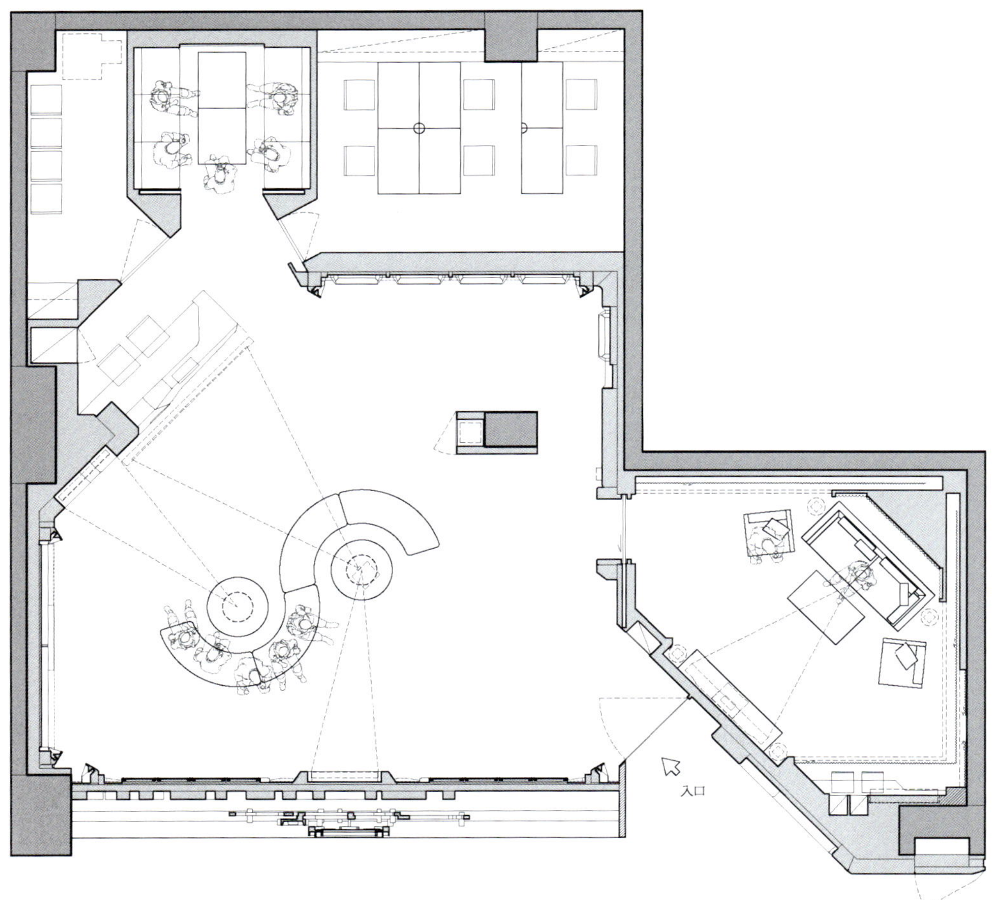

作为远雄建设进军大陆市场的试金石，设计将变形金刚的钢甲造型作为空间的主要元素，融入先进的科技互动装置，协同软硬件共构装设，实践了科技美宅的想望。

入口处引用机械表精密的作动原理，设计了一款名为“未来之瞳”的互动装置，借助相机光圈的感应启动装置，当造访者行经入口时，中间有如瞳孔的装置随即开展，配合特殊的声光效果，引领参观者探索未来生活的便利之道。

展示空间内部以代表企业形象的蓝色为基底，进门之后便能看见企业标志，四周透过金刚盔甲造型的立体模板展示概念，红色活动圈椅组合出S形休息区，顺势引导空间动线。设计在考虑空间机能的同时，导入风格美学，运用丰富的互动视觉元素，营造出空间与人文的灵活对话。

遠雄
Farglory

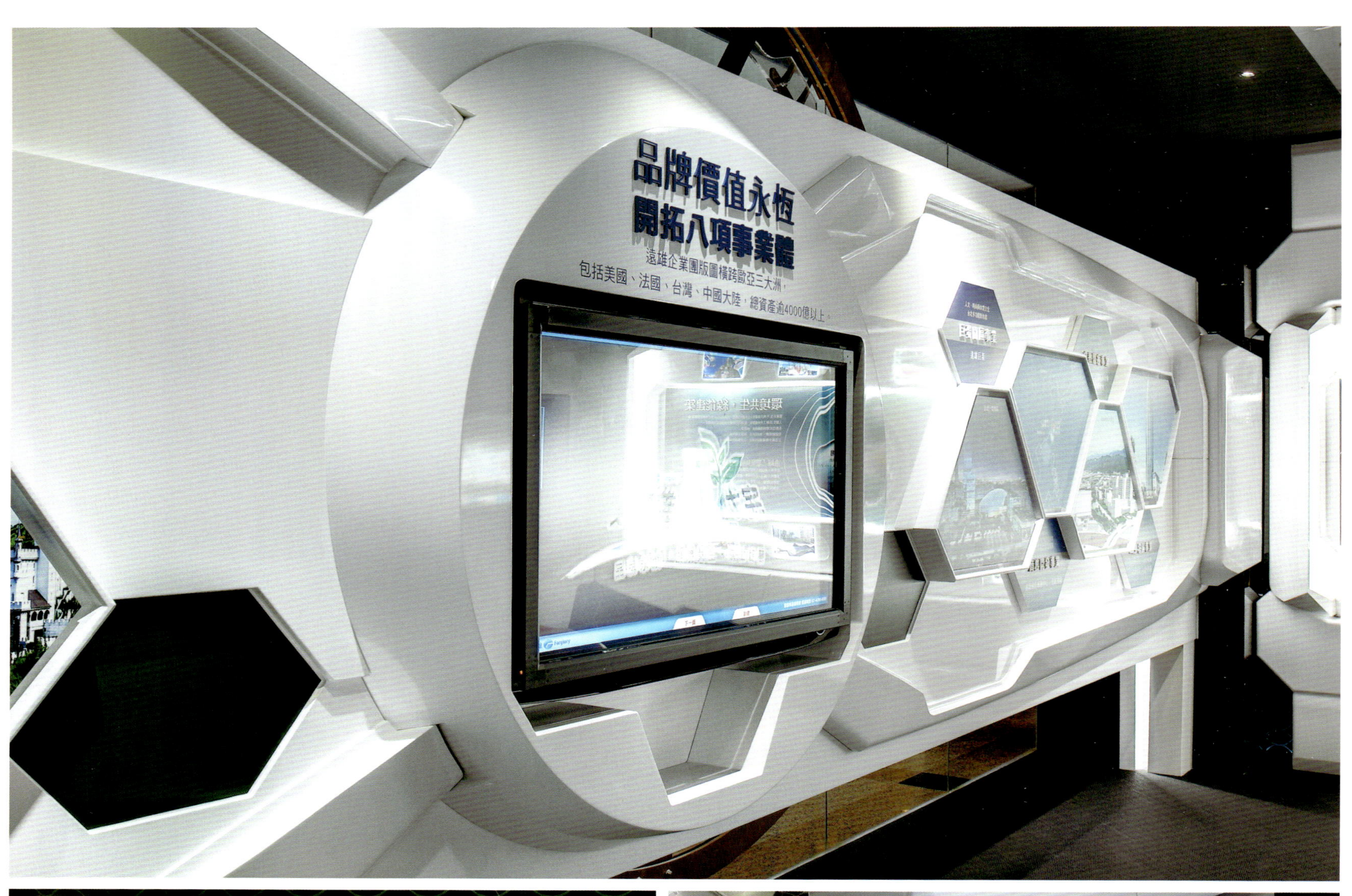
品牌價值永恆
開拓八項事業體
遠雄企業團版圖橫跨歐亞三大洲，
包括美國、法國、台灣、中國大陸，總資產逾4000億以上。

數位光纖家庭
D.F.H DIGITAL FIBER HOME
數位智慧的領先指標
遠雄數位12標章
建築
3C電器
傢具
傳播
機電
保全
通訊
醫療
娛樂
金融
服務
門禁

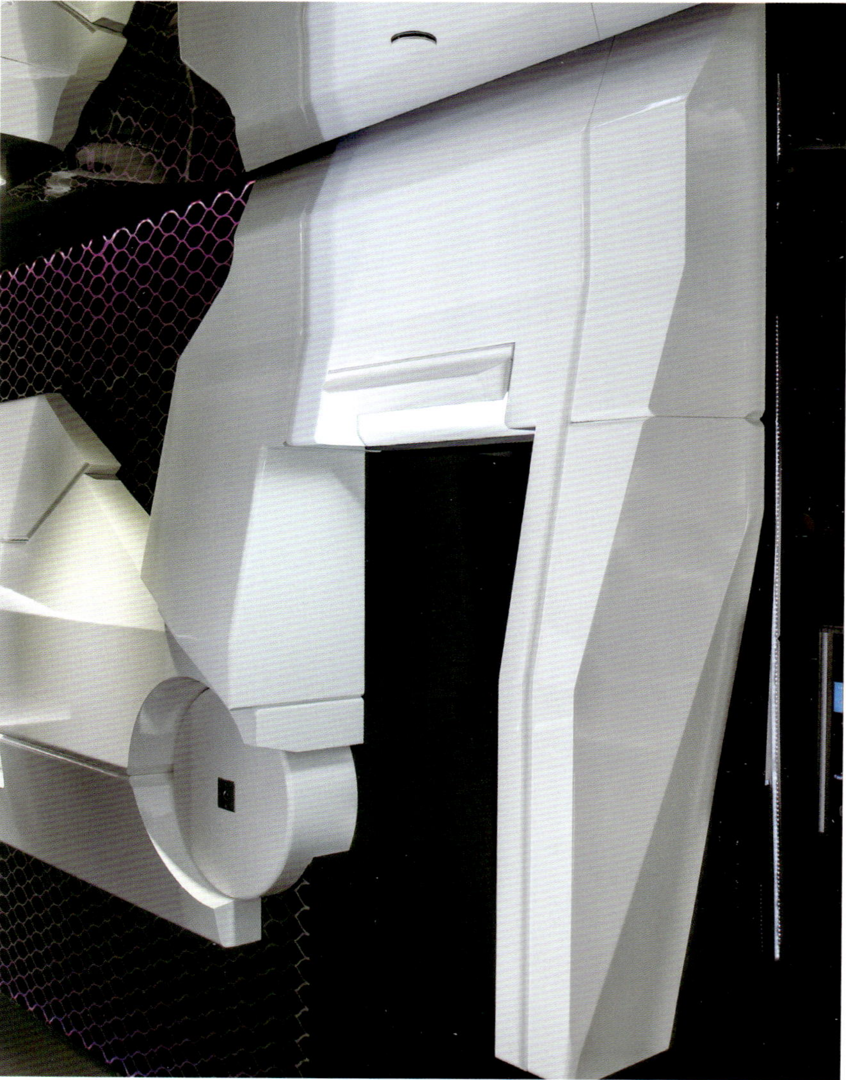

RICKY WONG
黄 志 达

黄志达设计师有限公司　设计总监

出生于中国香港知名的家具世家，受家族企业影响，孩提时期就接触了家具工艺和室内装饰，并开始了“设计创作”之路。自香港理工大学室内设计专业毕业后，他又选择了美国威斯康辛国际大学建筑学继续深造。1996 年黄志达先生在香港创立了自己的公司，开始为各类高端客户提供专业的设计服务，其独特的构思和巧妙的设计手法得到了广泛的认可。1998 年黄志达先生在深圳成立分公司，业务范围不断扩大。

在黄志达先生创作的众多享誉业界的作品中，有商业中心、酒店、高档会所、餐饮娱乐空间，也有不少名门望族的私人豪宅，项目更是遍布中国香港、北京、上海、广州、深圳、杭州、西安以及东南亚等地区。这些成功之作也为他赢得了 Gold Key Award、APIDA、亚洲最具设计影响力等国内外大奖。

经常行走于不同的城市和国家的他，善于细腻地观察和体验，研究不同的生活模式与人的需求，并将这些融入空间创作中。所以他的作品总是极具个性与意境，能为客户带去最大的利益和生活享受。其在设计上独到的见解也引起了业内外的关注，曾多次应邀做学术演讲，大量作品被各大媒体选登。黄志达先生出版过多本个人作品集，近年来推出的《安毕恩斯》及《安毕恩斯 II》更是为人们带来了不同空间的美学感受，不仅让设计界多出一个新的名词，还让越来越多的人去关注空间对人的影响。

Green Windsor Mountain Club

项目地址
东莞樟木头

设计师
黄志达

设计公司
黄志达设计师有限公司

绿茵温莎堡山顶会所

The project is located on the peak of a mountain in Dongguan, away from the hustle and bustle of the city, presenting a natural, open and breathing space for you. The strong sense of line is integrated with modern Southeast Asian technique, natural materials, simple geometric shapes, hollowed design and surrounding water landscape, so you can enjoy a long-lost leisurely feeling and close to nature.

The whole club is a fan-shaped two-floor structure, and large area of glass curtain wall features a sight without any barrier. The upper floor is completely open, in addition to enjoying the panoramic view at the foot of the mountain; you can enjoy an endless starry sky at night. The bar is made of black marble, coupled with several groups of irregular geometric modeling black rattan sofa on both sides, echoing with the pure white mats and cushions, filled with leisurely mood of Zen. The slant round tables hide surprises in the simple shape. You can sit down to enjoy coffee, to listen to the wonderful music, to share the lemon on plate, and to look at verdant plants… then, you are enjoying the most pleasant relaxing mentally and physically.

位于东莞山顶的会所，远离城市的喧嚣，为人们呈现了一个自然、开放、会呼吸的空间。超强的线条感融入现代东南亚的手法、天然的材质、简洁的几何造型，还有中空及四周环绕的水景，让人体会久违的悠然自得、亲近自然。

整个会所呈上下两层的扇形分布，大面积的玻璃幕墙让视线没有任何阻隔。上层完全是露天的，除了将山下的风景尽收眼底外，还能坐享繁星闪烁下的无边夜空。吧台以黑色的大理石打造，两旁分布着几组不规则几何造型的黑色藤艺沙发，呼应着纯白色的垫子和抱枕，禅意悠然。圆形的几台是倾斜的，简单中蕴藏着惊喜。坐下来享受咖啡的醇香，聆听美妙的琴声，分享托盘上的柠檬，观赏嫩绿的植物……让身心得到最愉悦的放松。

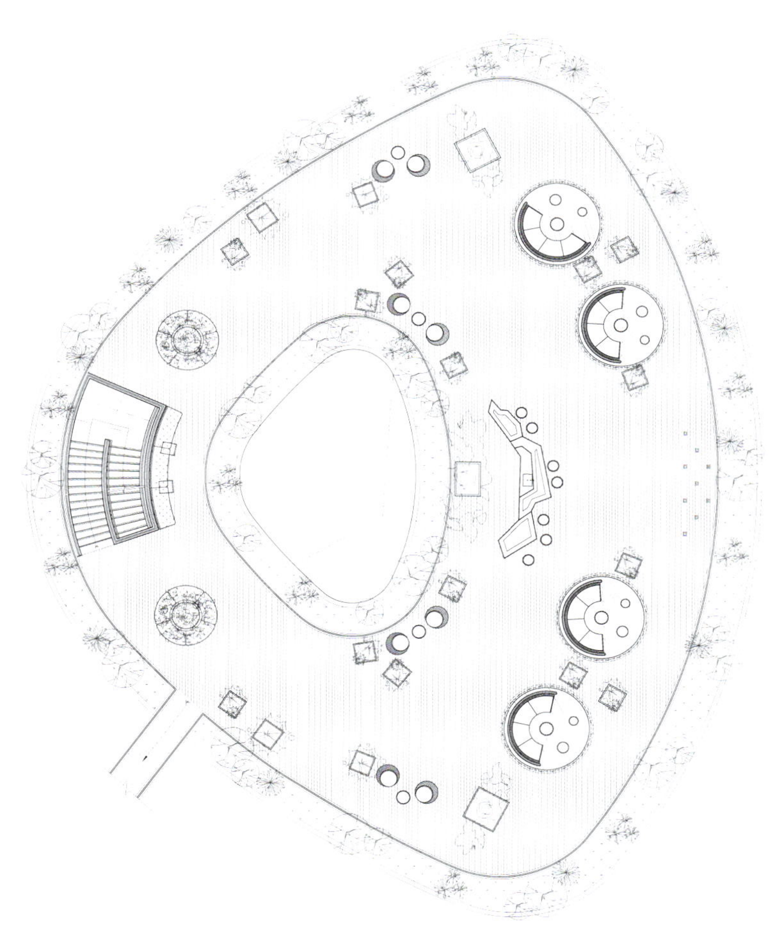

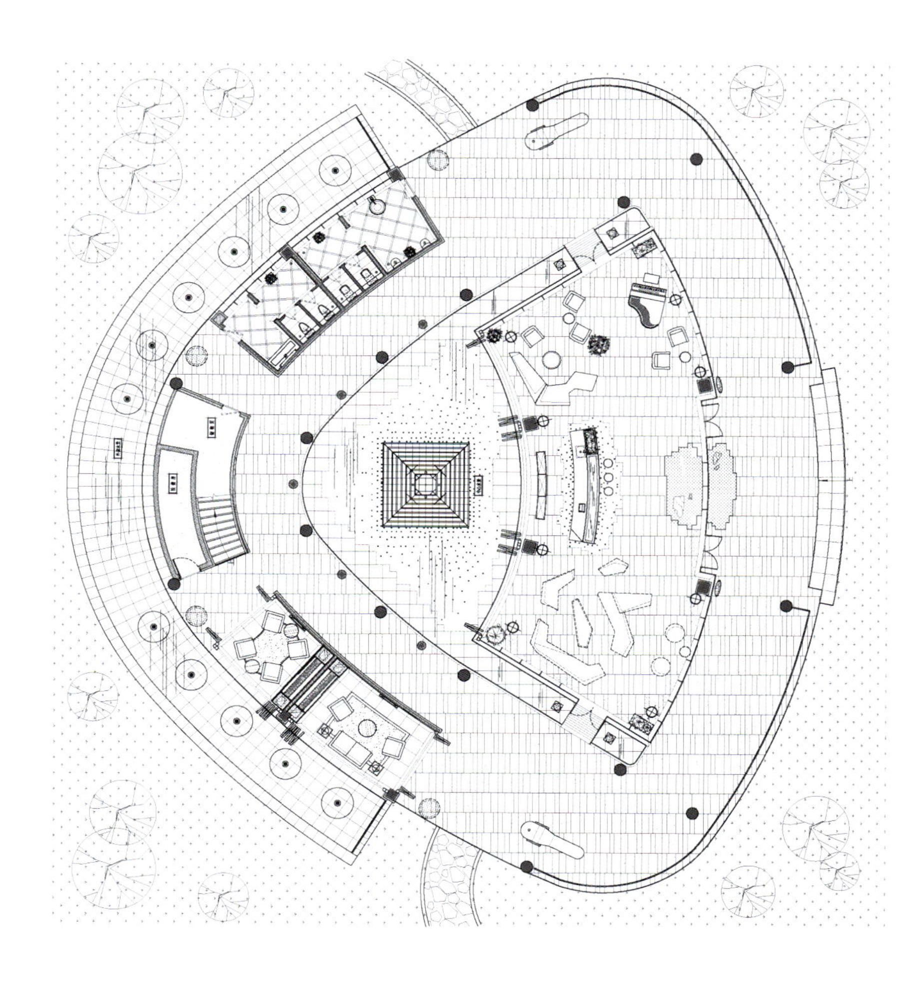

北区登山口
南区登山口
网球场
万象广场
篮球场
顶公园主入口
迎宾台
水云间

Kingswan Golf Club

项目地址
江苏常州

设计师
黄志达

面积
4880 平方米

设计公司
黄志达设计师有限公司

金沙湾高尔夫会所

Kingswan Golf Club is located in the endless green mountains, embracing the ups and downs of rolling terrain, and listening to the singing of birds. The designer creates an open and relaxing atmosphere, integrating the cultural connotation of golf with the extension of new Asian style to interpret the elegance and charm of golf.

Walking through the long corridor in front of the water landscape area, you will entrance the reception hall, which is also a golf souvenir shop, where the ceiling remains the original structure, but the reception table is made up of rough stone. On the left, a round area is skillfully designed to be a golf museum. The spiral staircase in the middle is surrounded by a two-storey high octagonal pavilion, and various kinds of bird models are hanging in the central hollowed place to outline a scene of "hundreds of birds paying homage to a phoenix" .

On the other side of the pavilion is the VIP area for ladies and gentlemen, where the wall and ground are soft light colors, and the furniture is darker wood color, decorated with selected hanging pictures, lively and interesting animal-shaped ornaments and gauze curtain imported from Turkey, to endow you with wonderful interactions and high-quality experience.

餐厅
Restaurant

金沙湾高尔夫会所位于一望无垠的绿野山地间，拥抱起伏绵延的地形，倾听清脆的鸟鸣。设计营造出开阔轻松的氛围，将高尔夫的文化内涵与新亚洲格调的外延相统一，演绎出高尔夫运动的优雅的魅力。

穿过水景区上方的长廊来到接待大堂，也是高尔夫纪念品商店，此处，天花保持了原结构，接待台则由粗犷的石块打造。往左走，圆形的区域被巧妙地打造成高尔夫博物馆。中间的旋转楼梯被两层高的八角亭包围，设计师在亭子中空的位置挂上了各种各样的鸟的造型，勾勒出“百鸟朝凤”的画面。

亭子另一侧是男女贵宾区，墙体和地面是柔和的浅色调，家具则是较深的木色，由精心挑选的挂画、生动有趣的动物摆件和从土耳其引进的纱帘点缀，带给人美妙的互动和高品质的体验。

HANK M.CHAO
赵 牧 桓

牧桓设计 + 灯光设计研究室　设计总监

2005 年 成立牧桓设计 + 灯光设计研究室
2002 年 成立上海牧桓 + 灯光设计研究室
1997 年 成立牧筑空间暨造型研究室（中国台湾）
1994 ~ 1995 年 Klodagh Designs, NY USA
1991 ~ 1995 年 纽约室内设计学院
2001 年 中国建筑学会室内设计会员
1998 年 中国台湾室内设计协会会员
1997 年 美国 IES 灯光设计协会会员

主要完成作品
2012 年
MoHen Design International 上海总部办公室
Tender Luxury 售楼中心

2010 年
上海仁恒河滨花园 Yanlord Riverside Garden
Luminaire 上海总部办公室
Apartment No.10 样板房

2009 年
现代宫廷 Palace in Modernity
卢湾海珀日晖售楼处 Twinkling Space
上海墅外 Shuwai

2008 年
百大集团样板房设计 Natural Organic Space
[illegible]样板房设计 The Chian Alley Memories
[illegible]iva 上海旗舰店
[illegible]ounge

[illegible]
蛋蛋屋快餐上海连锁店
上海永嘉路太原别墅
上海 Parkside 意大利餐厅

Angular Momentum

项目地址
上海

设计公司
牧桓设计 + 灯光设计研究室

设计师
赵牧桓

主要用材
密度板、冷烤漆、不锈钢、环氧树脂、玻璃

参与设计
赵玉玲、胡昕岳

摄影
周宇贤

Angular Momentum

The design idea of the project is to seek the moving kinetic energy and rhythm of the space through vectors, and the unequal two-dimensional plan forms three-dimensional objects to create agglomerative space kinetic energy. The plan blocks of different vectors are mutually linked and twisted while shifting and overturning on the axis, making walls static pictures which seem to move, thereby the space has a sense of movement. This is a visual implication, but it changes man's real perception, and the space features a different feel as well.

The designers make use of the partition walls in the seating area to divide the circulation at the entrance, and make it the first curtain at the entrance of the space, to divide the large private working area from the open public area. The small and medium-sized conference rooms and billiards entertainment rooms are located on both sides. The partition walls between the conference rooms highlight the working attribute of office through linear carved glass. The reference room is open without obvious segmentation, linked by a walkway in the middle, which can be used as a small communication area for temporary discussion, and the walkway directly leads to the head office and then splits to lead to entertainment room and gym.

设计部
DESIGN DEPT.

设计部
DESIGN DEPT.

透过向量寻找空间流动的动能和律动是本案的设计思路，不等分的二维平面进而构成三维的量体，形成一种凝结的空间动能。不同向量的平面切块互相交汇连结并在轴线上扭动翻转，让墙面成了处于运动状态的静止画面，空间也因此有了流动感，这是一个视觉暗示，但却改变了人的实质感知，空间也有了不同的趣味。

设计师利用休息区的隔墙分流入口动线，并将其作为进入空间的首要门帘，划开私密的大工作区块与公共的开放区间。两侧分布中小型会议室和台球娱乐室。会议室的分隔墙以线性雕刻玻璃强调办公室的工作属性。资料室采用开放式设计，并无明显区隔，中间由一个长廊连接并可作为一个小型的临时讨论短会交流区，走道直通主管办公室并分流至娱乐健身房。

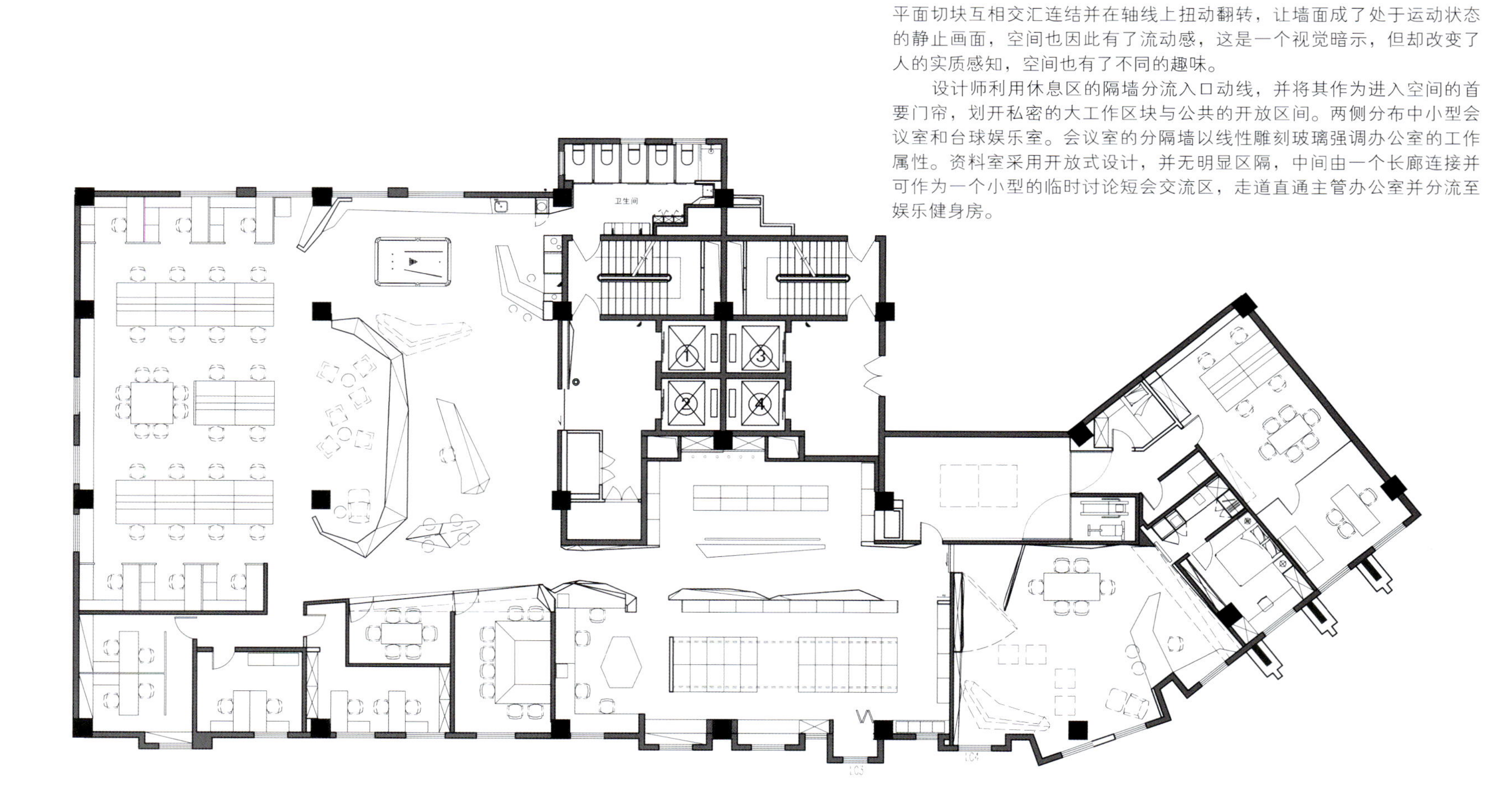

DESIGN DEPT.

图书室
LIBRARY
3450

301
朋友
FRIENDS
厂商
SUPPLIERS
MOHEN
?!
WRONG WAY

Go
3rd
Floor

Go
3rd
Floor

2400
1150

项目地址
上海

设计公司
牧桓设计 + 灯光设计研究室

设计师
赵牧桓

主要用材
橡木、黄洞石、玻璃、铁件、白杨木

参与设计
王颖建、胡昕岳

摄影
周宇贤

Tender Luxury

The project is located on the bank of Huangpu River in Pudong, Shanghai, surrounded by well-planned landscape supporting facilities.

The entrance deliberately emphasizes the depth of landscape, with water pool separating it from the back circulation leading to the washing room. The wall facade leading to the washing rooms features the pattern of flowing water to reflect water waves, visually allowing the interior "river" to communicate with the outdoor Huangpu River. The seating area is separated by bookcase hanging from the ceiling without touching the ground, so vision features a sense of penetration in some parts. In addition, the bar is designed with carving technique to form a contrast with the surrounding straight lines, and the white poplar woods served as background make the light and shadow looming, further enhancing the poetic sense. The walkway leading to the VIP private rooms is also projected with water waves on the ground, and the dynamic way endows the process into the private rooms with fun, and makes up the limit of non-interaction with people for the "hardware" design defects. The colors echo with the conceptual extravagance through a calm and low-key tone, and show a sedate magnificence through materials.

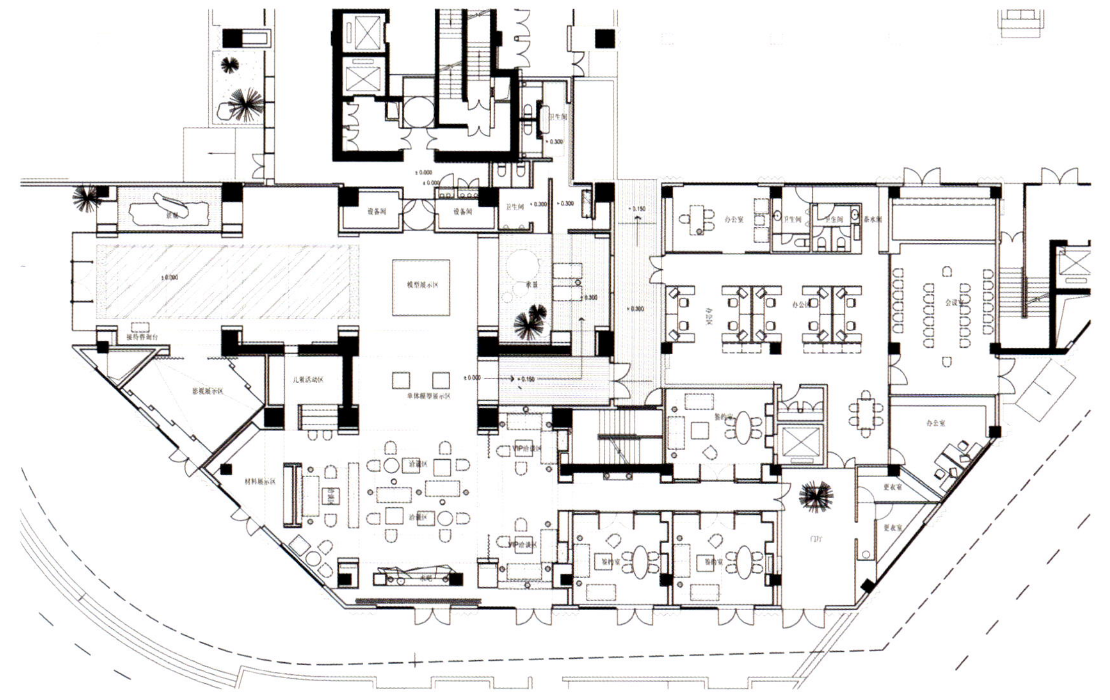

本案坐落在上海浦东的黄浦江边，周围有良好规划的景观配套设施。

在入口处刻意强调景深，并用水池隔开后方通往洗手间的动线。通往洗手间的墙立面以水波纹动态投影投射流动水的质感，让室内与室外的黄浦江有了视觉上的联系。座位区则利用从天花下来不到底的书柜作为分隔，让视觉有部分的穿透感。比外，以雕刻的手法处理吧台，与周遭的直线条做出对比，背景衬托白杨木的树林让光和树影若隐若现，更增加了诗意感。通往贵宾包厢的走道也同样在地面上投影水波，这种动态的方式为进入包厢的过程增添了趣味性，也弥补了设计局限于“硬体”而无法与人互动的缺陷。色彩上以沉稳、内敛的基调呼应概念上的铺张感，并透过材质表现出一种深沉的华丽。

KELLY HOPPEN

FRANK JIANG
姜 峰

J&A 姜峰室内设计有限公司　创始人及总设计师

他是目前中国室内设计行业唯一享受国务院特殊津贴的专家，现任中国室内装饰协会设计委副主任、中国建筑学会室内设计分会副会长等社会职务。并被聘任为中央美院、天津美院、鲁迅美院、广州美院的客座教授，从事中国商业室内设计发展方向的研究工作，发表了多篇有关购物中心设计的学术论文，引领着中国商业室内设计的发展方向。

2009 年，应在国际上有着重大影响的“伦敦艺术节”组委会的邀请，作为中国唯一的特邀演讲嘉宾进行了“室内设计对 SHOPPING MALL 商业价值的驱动”的演讲。受到了国内外的广泛关注。

姜峰先生在他 32 岁的时候便被评为“深圳市十大杰出青年”，37 岁时被评定为教授级高级建筑师，39 岁时荣获了“终身设计艺术成就奖”，并被评选为“中国建筑装饰设计领军人物以及全国有成就资深室内建筑师”，44 岁时被评为“深圳市福田区杰出人才”。

Dreamport Beijing

项目地址
北京

设计公司
J&A 姜峰室内设计有限公司

面积
200000 平方米

设计合作
贝诺建筑设计公司

设计师
姜峰、袁晓云、陈礼庆

主要用材
天然石材、人造石、瓷砖、艺术玻璃、铝板、不锈钢、乳胶漆

北京华润五彩城

Dreamport Beijing is a multi-functional space integrating with shopping, dining, entertainment and leisure, and sports. The concept of living mall not only creates an atmosphere of "living field" in the format, but also endows the building with affinity. The lighting effect of Dreamport is warm and soft with clear levels, creating a comfortable lighting environment. Every single piece of light and shadow is linked up appropriately, exuding a quiet and bright beauty of harmony. The developed commodity production and commercial competition attract customers, stimulate consumption, and lead shopping to transform from "requiring purchase" to "inspiring purchase" through constantly updating and changing. Therefore, you cannot ignore the importance of shopping environment. The purpose of the commercial lighting design is to create a comfortable, easy to watch, appealing and attractive lighting condition. Dreamport Beijing extends to different directions, creating a place integrating with shopping, entertainment, culture and leisure, attracting and meeting various needs of customers.

莎莎
THE NORTH FACE
hotwind
就在一层
sasa
CONVERSE

hotwind

2011-2012赛季

电梯上行

冰场
Ice Rink
CASIO
CASIO
G-SHOCK
Baby G
PRO TREK

北京华润五彩城拥有购物、餐饮、娱乐休闲、运动等多功能的空间。living mall 的概念，不仅在业态上为其营造了“生活场”的氛围，还使建筑充满亲和力。五彩城的照明效果温润柔和，层次分明，创造了一个舒适的光环境。每一道光与影都被恰如其分地衔接起来，散发出静谧与光明的和谐美。

发达的商品生产与商业相互竞争是以商品的不断更新变换来吸引顾客、刺激消费、引导购物从“所需购买”向“激发购买”转化的。因此，购物环境的作用不容忽视。而商业照明设计的目的就是营造一个舒适、易于观看、富于感染力和吸引力的光照环境。北京华润五彩城向多方位延伸，形成了一个购物、娱乐、文化、休憩相结合的场所，吸引并满足了顾客多方面的需求。

U-town Lifestyle Center

项目地址
北京

设计公司
J&A 姜峰室内设计有限公司

面积
110000 平方米

设计合作
香港创智建筑师有限公司

设计师
姜峰、袁晓云、陈礼庆

主要用材
天然石材、人造石材、瓷砖、复合木地板、木纹铝板、不锈钢、艺术玻璃、透光软膜、乳胶漆

北京悠唐生活广场

U-town Lifestyle Center is a comprehensive shopping plaza integrating with shopping, entertainment and leisure, and it is also the first mixed-use project in Beijing. The large one-stop comprehensive consumption center features the largest indoor center square, the most special delicacy resort, the most fashionable leisure world, and the most unique air SHOW site in Beijing.

The center is a typical neighborhood-style layout, where the whole commercial facilities are organized and planned in the neighborhood enclosed by outside streets. It adopts the way of pedestrian zone, and at the same time, it is connected with the outside traffic, convenient and independent. The comprehensive consuming layout of "shopping+dining+entertaining" provides a free enjoyment at the center of the city for business elites in Beijing.

In the whole exhibition space, the designers take all possible factors, striving for ingenuity in colors, lighting and decoration. In terms of layout, the designers personalize the way of exhibition, so that the whole space and process is integrated and perfected.

ZARA

热恋钻石商场

Superdry.

北京悠唐生活广场是一家集购物、娱乐、休闲于一体的综合性购物广场，同时也是北京首个城市综合体项目。一站式大型综合消费中心，拥有北京最大的室内中心广场、最丰富的特色餐饮美食总汇、最时尚的休闲世界、最具特色的空中 SHOW 场。

广场采用典型的街坊式布局，整体商业设施都在外围道路围合的街坊内组织规划，采用步行区的方式，同时又和外部交通相联系，便捷而独立，同时“购物 + 美食 + 娱乐”的综合性消费布局为北京的商务精英们提供了位于城市中心的自由享受。

设计师在整个展示空间中调动一切可能配合的因素，在色彩、照明、装饰手法上力求别出心裁，在布置方式上将展示陈列人性化，使整个空间和过程完整、完美。

Henan Sky Land Bay Club

项目地址
郑州

面积
443794.33 平方米

设计师
姜峰、袁晓云、陈礼庆

设计公司
J&A 姜峰室内设计有限公司

主要用材
灰镜、茶镜、地毯、圣地亚哥石材、黑洞石、灰木纹石材、白橡木

河南天地湾禧苑会所

Sky Land Bay Club integrates club, residential and commercial houses into a whole, relying on the high-quality platform of Sky Land Bay International composite living area and the new Asian Villa of International Noble Residence Phase Ⅰ, organically combining the international quality and Oriental context, bustling city and nature ecology, to highlight the strong new Asian style.

The whole top is designed with gentle slope roof and large cornices, to highlight the strong charm of the Han and Tang Dynasties, harmoniously responding to the modern and simple texture of the vertical facade. Large area of stone is introduced into the design, coupled with building facades in reddish-brown bricks to outline a changing building texture, highlighting the elegant texture and featuring heavy cultural heritage.

In the club, the design of ceiling and window lattices introduces the concept of China's national flower peony, with the hope of expressing China's cultural temperament and mood through modern technique and showing a neo-classical space. The aspects of structure, color and lighting are particular about the changes of levels and textures, highlighting the depth to achieve the space effect.

天地湾禧苑集会所、住宅、商用等多功能于一体，依托天地湾国际复合生活区优质平台、国际高尚居住板块一期新亚洲主义院墅，将国际品质与东方文脉、城市繁华与自然生态有机结合，彰显浓郁的新亚洲主义风情。

顶部整体采用缓坡屋顶大挑檐设计，彰显浓重的汉唐风韵，与纵向立面现代、简洁的质感和谐相依。大面积的石材运用，与红褐色面砖相间的建筑立面，勾勒出富于变幻的建筑肌理，突显尊贵质感的同时，更添厚重的文化底蕴。

会所在天花、窗棂的设计中融入中国国花牡丹花的概念，希望以现代手法来表现出中国的人文气质和意境，展现新古典的空间。在结构、色彩、照明等方面则讲究层次、肌理变化，凸显出深度，达到丰富的空间效果。

Tiens R & D Quality Inspection Center

项目地址
天津

面积
8800 平方米

设计师
姜峰、袁晓云、陈礼庆

设计公司
J&A 姜峰室内设计有限公司

主要用材
地灰麻、安提克灰A级、雅士白、青花纹、白色人造石、橡木

天狮研发质检中心

Tiens R&D Quality Inspection Center is a mixed-use office with main functions of research and development and quality inspection of the biomedical health care products. The whole building is divided into three parts, R&D center on the third floor, quality inspection center on the fifth floors and multi-functional hall, and R&D center and quality inspection center are connect through a oval lobby.

Combined with the features of R&D Quality Inspection Center, the project refines the DNA's hyperbolic shape of double helix structure and makes it the main design element, showing the symbolic meaning of the center as the biomedical high-tech core striving for creation and nurture. As the core of the whole building, the lobby connects the R&D Center and Quality Inspection Center, playing a role of bridge in the space circulation. The designers change the original straight gallery bridge into hyperbolic shape, echoing with the DNA designing concept, and as the origin of the curved shapes, it diffuses and extends to other spaces, which organizes out the movement and circulation of the whole space, vividly reflecting the fun and variability of the building.

天狮研发质检中心是以生命医学保健品的研制开发和质量检测为主要功能的综合办公楼。整个建筑分为研发中心（三层）、质检中心（五层）、多功能厅三部分，研发中心和质检中心通过椭圆形的大堂相连接。

本案结合研发质检中心的功能特性，提取 DNA 双螺旋结构的双曲线为主要设计元素，蕴含着研发质检中心作为创造、孕育生命医学高新技术核心力量的象征意义。大堂作为整栋建筑的核心，连接着研发中心和质检中心，在空间流动上起着桥梁作用，设计师将原有的直廊桥改造为双曲线造型，呼应了 DNA 的设计理念，同时作为曲线造型的原点发散、延伸到其他空间，从而组织出整个空间的动向与流线，生动地体现了空间本身构筑的趣味性与多变性。

CTS Tycoon (Shenzhen) Golf Club

项目地址
深圳

面积
14000 平方米

设计师
姜峰

参与设计
袁晓云、陈礼庆、黄日金

设计公司
J&A 姜峰室内设计有限公司

主要用材
贝沙金石材、皇家奶白石材、肌理涂料

深圳港中旅聚豪高尔夫球会所

CTS Tycoon (Shenzhen) Golf Club is aimed at returning to tradition and close to nature, allowing customers to experience pure Scottish highland-style architecture and the profound golf culture. Through typical arch shapes, bare wooden beams, rough rubbles and plaid cloth, the design leads customers to go away from the bustling city and to feel the Scottish flavor.

In the design of the lobby, the designers respect the inherent architectural structure to retain the dome design and introduce decoration of colored glaze glass, so that the space is more close to nature. The space handling technique is concise, paying attention to the primary and secondary, arranging them with appropriate proportion. The lighting dome, curved staircase and reception desk are the focus of the space design, colored glaze glass of lead mosaic, antique materials and wrought iron carvings make the space more detailed, make it real and exquisite, so customers will feel like in the Scottish castle, enjoying the fun from golf.

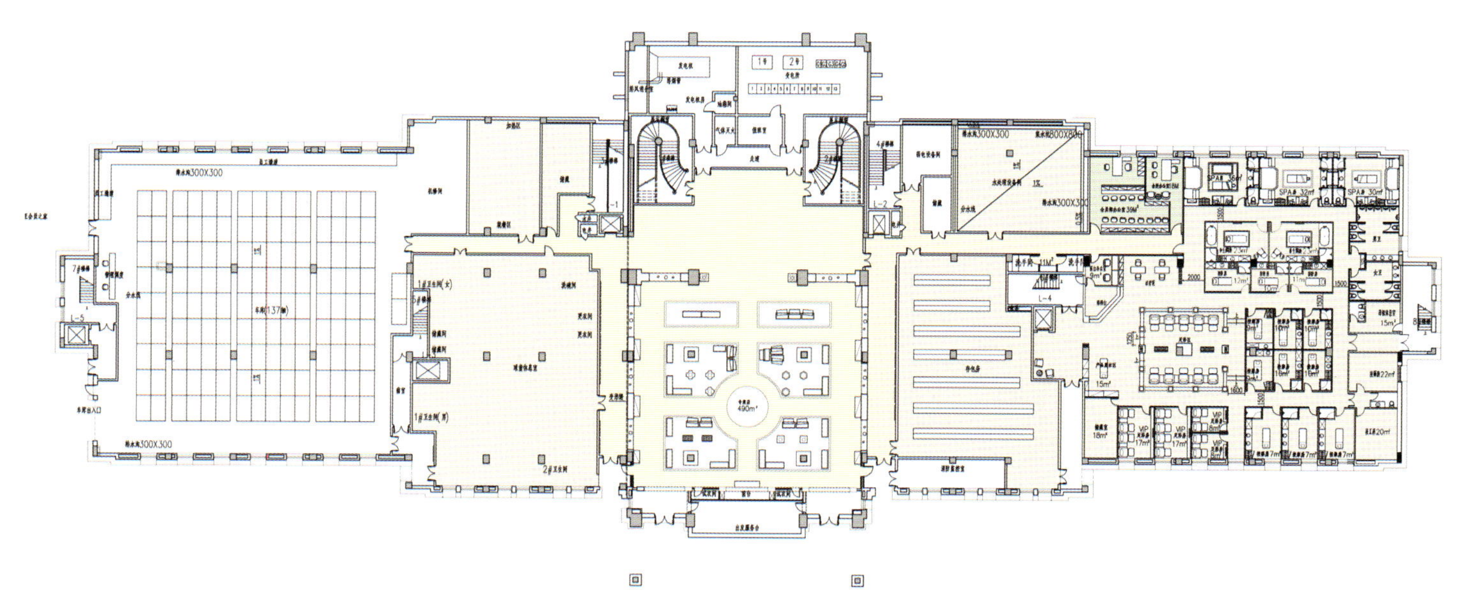

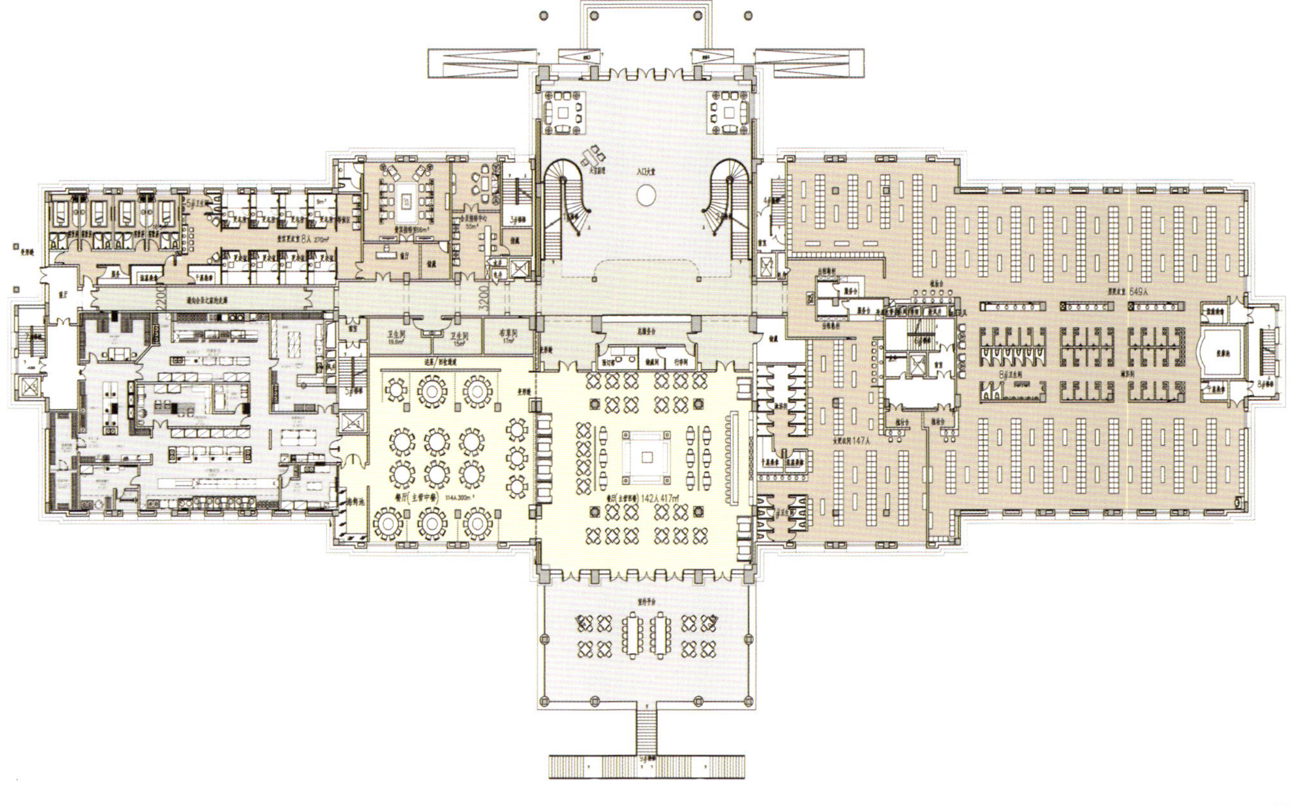

聚豪高尔夫球会所的宗旨是回归传统、亲近自然，让人们在这里体验到纯正的苏格兰高地风格建筑及深厚的高尔夫文化。通过典型的拱形造型、裸露的实木横梁、粗犷的毛石和格子布艺，让人们远离城市喧嚣，感受苏格兰风情。

大堂的设计中，设计师尊重其固有的建筑结构，保留了穹顶设计，并用彩釉玻璃装饰，使空间更贴近自然。空间处理手法简练，注重主次、比例。采光穹顶、弧形楼梯和服务台是空间设计的重点，铅条镶嵌的彩釉玻璃、仿古面材料、铁艺雕花使空间更具细节感，真实而细致，让人仿佛身临苏格兰城堡，享受高尔夫带来的乐趣。

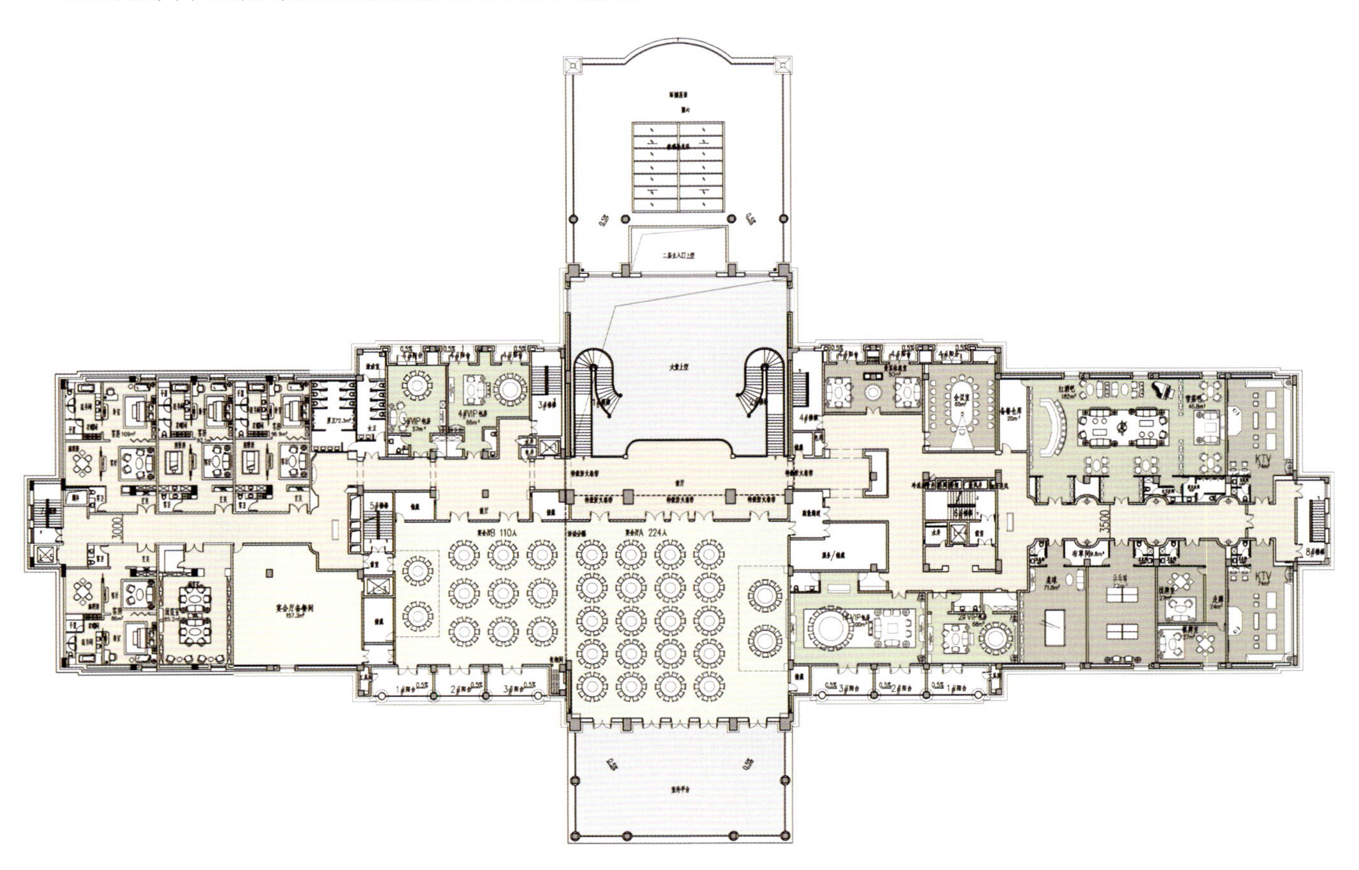

GAME OF FATE
RILEY

Tiens Education Training Center

项目地址
天津

面积
28000 平方米

设计师
姜峰、袁晓云、陈礼庆

设计公司
J&A 姜峰室内设计有限公司

主要用材
白麻、灰麻、木纹铝板、灰橡木饰面、枫木吸音板、拉丝纹不锈钢、白色铝板

天狮教育培训中心

The project is a public building mainly for meeting and training, and the interior space includes two multi-functional 400-seat halls, which can be separated, a terraced training hall which can hold 980 people and a round training hall with 3828 seats; the supplementary facilities are showing gallery, VIP reception room, VIP conference room, prayer room, large and small classrooms, dressing rooms and others.

The designing idea of the project changes from the "pyramid" management mode of scientific team representing the corporate, to the understanding and interpretation of the corporate leading the industry, then to the thinking of "diamond"; from the interior, it changes from looking for the designing language suit for the building, to the refinement of the combination of triangular elements and curved contour, and finally it goes to reflect the golden rule of plan composition. The interior space design refines a healthy, environmental-protection and people-centered designing idea, introduces international mode-digitalized processing techniques, and rationally organizes points, lines and surfaces. It highlights different space regions through the changes of different materials and lighting, striving to create an international humanized meeting and training space for Tiens Group.

本项目是一个以会议和培训为主的公共建筑，室内空间主要包括两个可分隔的400人多功能厅、一个980人的台阶培训厅及一个3828人的圆形培训厅；辅助设施有展示通廊、贵宾接待室、贵宾会议室、祷告室、大教室、小教室及化妆室等。

项目的设计思想体现了从象征企业“金字塔”的科学团队管理模式、企业引领行业先锋的理解和诠释，到“钻石”的思维转化；从室内寻找呼应建筑本身的设计语言，到提炼三角形元素与弧形轮廓的组合，此外，还体现了平面构图的黄金法则。室内空间设计提炼健康、环保、以人为本的设计思想，采用国际化、模数化的处理手法，合理组织点、线、面，通过不同材质的变化及照明来强调空间的区域性，力图为天狮集团打造一个国际化、人性化的会议及培训空间。

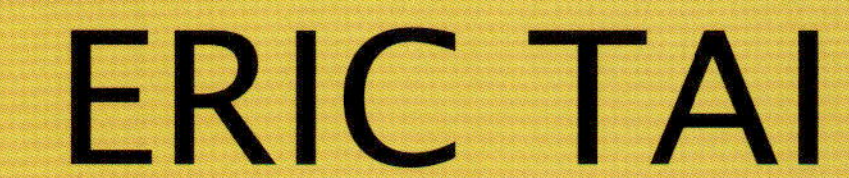

ERIC TAI
戴　勇

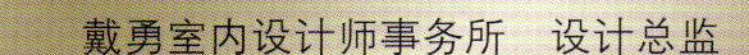

戴勇室内设计师事务所　设计总监

戴勇室内设计师事务所（以下简称戴勇设计）以及卡萨艺术品有限公司由中国著名室内设计师Eric Tai戴勇先生创立，专业从事会所酒店、销售中心、别墅样板房室内设计及艺术陈设服务，为顶级品牌室内设计机构之一。戴勇设计拥有极其丰富的设计经验及设计知识，一贯以卓越的设计管理及睿智创意服务客户；通过对国际设计潮流和未来设计趋势的前瞻性掌控，不断影响和提升室内设计行业的水准。

逾十余年，戴勇设计通过对各种风格的娴熟驾驭、功能需求和艺术美感上的精准平衡及对视觉效果和舒适自然感的把握，把原创优雅、尊贵品位的设计理念渗透到每个设计中，通过设计为客户创造了极大的价值，并取得了商业上的更大成功，从而也为戴勇设计赢得了更多服务客户的机会。

近年来，中国室内设计行业的飞速发展，也让更多的中国本土设计师成长起来，并有机会在国际舞台上展现中国的风采。2009年"中国十大设计师"戴勇先生又获选英国Andrew Martin室内设计奥斯卡全球优秀室内设计师，成为深圳首位获此殊荣的设计师。从戴勇设计创立至今，在亚洲先后为近两百个项目提供了极富创意及人文特色的室内设计及陈设设计服务工作。

戴勇设计为客户提供完整的设计服务，涵盖室内设计、陈设设计、陈设总包及监理服务，将具卓越创意的设计与精湛成熟的工艺完美结合，配合极具人文品位的陈设艺术，力图尽善尽美。在项目管理实施过程中，提供完善的监理及跟踪服务，从而达到控制成本和工期的管理目标，使客户能集中精力于其核心业务。

戴勇设计近十年来撰写出版了《逸境》《时尚奢华样板》《陈设生活智慧》《城市酒店与度假酒店设计》《极上雅境》《中国式优雅》（即将出版）等多本设计专著，为室内设计行业的发展带来了不可低估的贡献，是戴勇设计成为顶尖设计品牌的关键所在。

Shouyu Modern Show Flat

项目地址
深圳

面积
90 平方米

设计师
戴勇

设计公司
Eric Tai Design Co.,LTD 戴勇室内设计师事务所

软装设计
戴勇室内设计师事务所 & 深圳市卡萨艺术品有限公司

主要用材
希拉克木纹云石、银灰洞云石、透光云石、灰影木饰面、墙纸、清镜

摄影
江国增

首誉现代样板房

浅色的灰影木饰面围合出清新雅致的居家氛围，让只有 90 平方米的两层小复式，一样拥有丰富的内容和精致的细节，折射出生活的方方面面。

镜面与灰影木饰面的搭配，在简约中散发出清新时尚的格调。透着暖黄色灯光的云石台面，加上沙发背景墙从顶上垂下来的镜面不锈钢条，让人眼前一亮。精致优雅的饰品和摆设，哪怕只是一盆花艺，一把汤匙，一只酒杯，都令人爱不释手，让人不由自主地便感受到屋主的生活情趣。主人房的背景墙、软包的分缝、一盏水晶吊灯和一幅精选的挂画便成就了一组和谐的构图，简洁却舒适。

The light gray timber veneers enclose out a fresh and elegant home atmosphere. Although it is a small duplex structure of 90 square meters, it features rich content and exquisite details, reflecting all aspects of life.

The collocation of mirror and gray timber veneers expresses a fresh and fashionable style in the simple design. The marble countertop reflecting warm yellow light, coupled with the mirror stainless steel strips hanging down from the top of the sofa in the background wall, lightens up your eyes. The exquisite and elegant decorations and ornaments, even a pot of flower, a spoon, or a chalice, will make you love them and involuntarily feel the life style of the owner. In the master bedroom, the background wall is designed with soft packed seams, a crystal droplight and a selection of wall pictures, creating a group of harmonious composition, which is simple but comfortable.

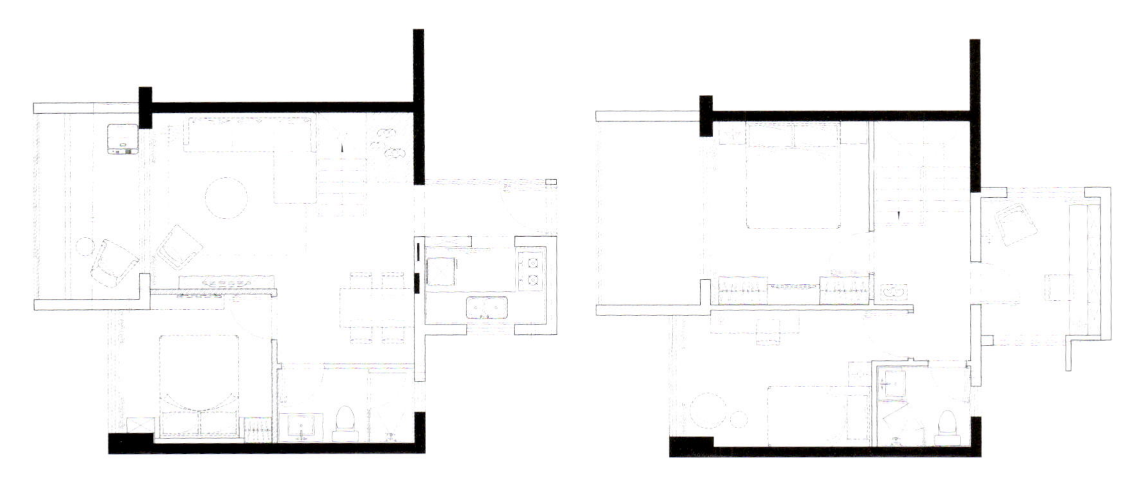

Runjin Blue Cartier Club

项目地址
莱州

面积
1200 平方米

设计师
戴勇

设计公司
Eric Tai Design Co.,LTD 戴勇室内设计师事务所

软装设计
戴勇室内设计师事务所 & 深圳市卡萨艺术品有限公司

主要用材
世纪米黄云石、浪漫紫荆云石、法国流金云石、墙纸、皮革硬包、黑檀木饰面

摄影
江国增

润锦蔚蓝卡地亚会所

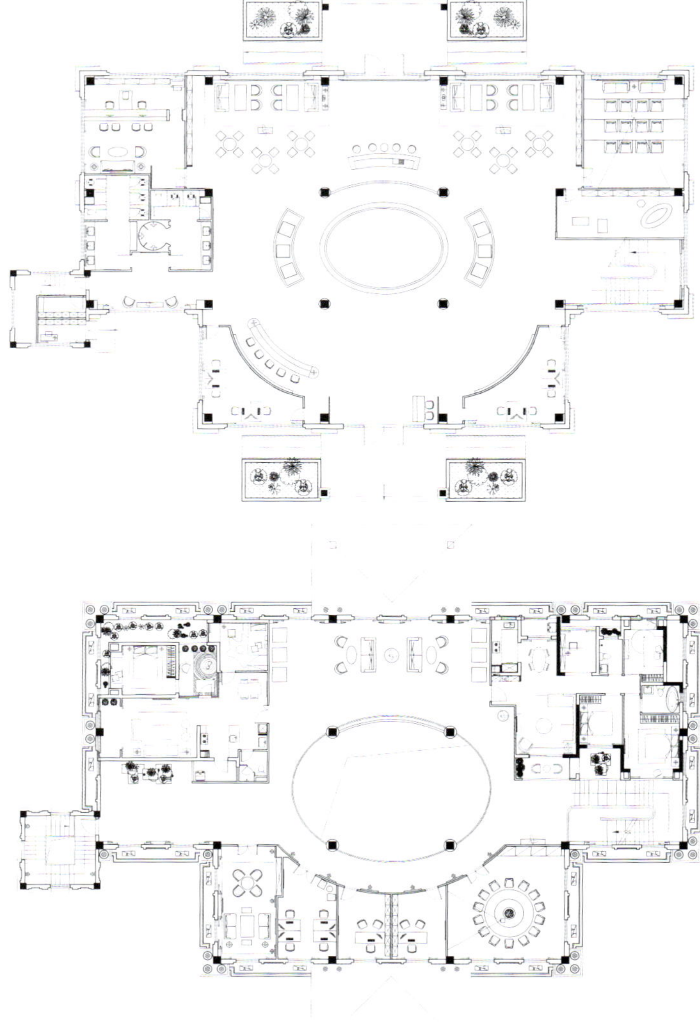

设计师彻底颠覆了原建筑格局的沉闷、闭塞，采用中轴对称的平面布局，展现出蔚蓝卡地亚会所新古典主义风格的高贵稳重；并且将上下两层打通，在展厅中央形成一个开放的椭圆形中庭空间；四根圆柱雄浑粗壮，二层天花通过一圈圈椭圆形的灯带使中庭显得更加高耸。一改旧貌，大气优雅。

椭圆形的平面布局打破了原先四方、保守的空间格局，同时又赋予了空间新的设计概念。中庭一圈圈的天花和地面上的圆形相呼应，围绕着中间椭圆形的沙盘，像一圈圈扩散开来的涟漪。椭圆形作为该项目独特的设计语言被运用到多个地方，加上精心挑选和摆设的家具及饰品，展现出新古典主义的尊贵姿容，同时又兼具优雅时尚的现代感，营造出居家的轻松氛围。

The designer overturns the original rigid and inconvenient architectural pattern, introduces axial symmetry layout to show the noble and calm neo-classicalism style in Blue Cartier Club. The upper and lower floors are interlinked to form an open oval atrium in the center, where four circular columns are forceful and stout, and the ceiling on the second floor is designed with circles of oval light beams to make the atrium even higher. The original appearance is changed to appear a new look which is generous and elegant.

The oval layout breaks the original conservative square pattern, and also endows the space with a new designing concept. The circles on the ceiling of the atrium echo with the circle on the ground, surrounding the oval sand table in the middle, which look like circles of ripples spreading out. As the unique designing language, oval is applied in more places, coupled with carefully selected and placed furniture and furnishings to show the noble appearance of neo-classicalism, at the same time, it features elegant and fashionable modern sense, creating an easy home atmosphere.

THE
BODY

合肥许建国建筑室内装饰设计有限公司 设计总监

安徽省建筑工业大学环境艺术设计专业
进修于中央工艺美术学院室内设计大师研修班
武汉艺术学院设计艺术学硕士研究生班毕业
CIID 中国建筑学会室内设计分会会员
国家注册高级室内建筑师
中国建筑室内环境艺术专业高级讲师
中国美术家协会合肥分会会员

Beijing Shouzhou Hotel

项目地址
北京

面积
16000 平方米

设计师
许建国

参与设计
陈涛、欧阳坤、程迎亚

设计公司
合肥许建国建筑室内装饰设计有限公司

主要用材
意大利木纹石、水曲柳肌理板、仿古砖、原木、皮革

摄影
吴辉

北京寿州大饭店

Beijing Shouzhou Hotel takes Chinese style as the main clue of designing, aiming at conveying Chinese flavor, to show Chinese cultural deposit and artistic charm. The project features a large area and an open space, fully showing the unique charm and temperament of Chinese style; coupled with the regional culture to perfectly integrate with the surrounding environment, further showing the delight of "little breeze and warm woods, with lotus overflowing everywhere".

Generally, the hotel is decorated with Chinese style, and the design of guestrooms introduces other styles, with exotic elements and accessories to fully show the inclusive and artistic space. The specific and clear separation of functions perfectly balances the sense of space. The decorative design fully reflects the characteristics and charm of Chinese architecture, where the bricks, gray tiles, antique tiles, antique furniture, screens and wooden shelves are combined with modern materials to create a low-key and elegant frame, coupled with classical Chinese elements, such as ceramics, paintings, wooden carvings, stone carvings, bamboo and other embellishments to enrich the space, and life therefore is more successful.

祥和百年

酒店设施名录
HOTEL DIRECTOR
-1F SPA
LUNCH BOX
1F
LOBBY TOTALSTATION
自助餐厅
BUFFETRESTAURANT
商务中心
BUSINESSCENTRE
特产中心
SPECIALTYCENTER
2F 客房
GUESTROOM
3F
GUESTROOM LEISUREAREA

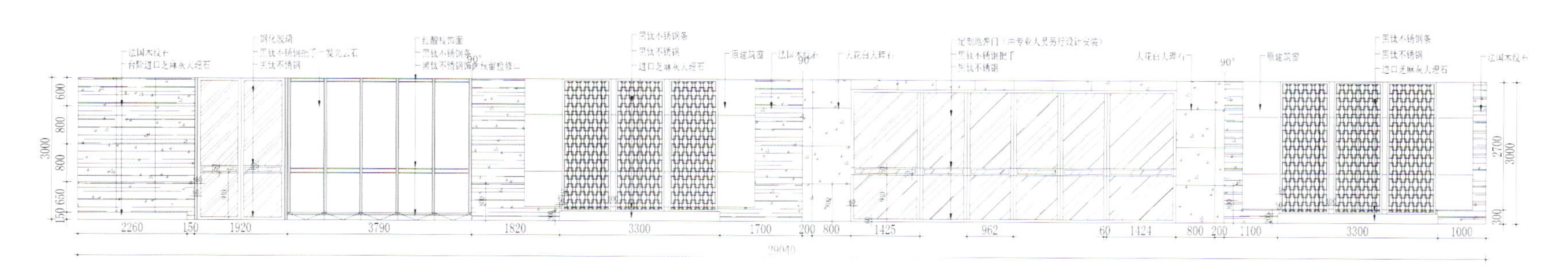

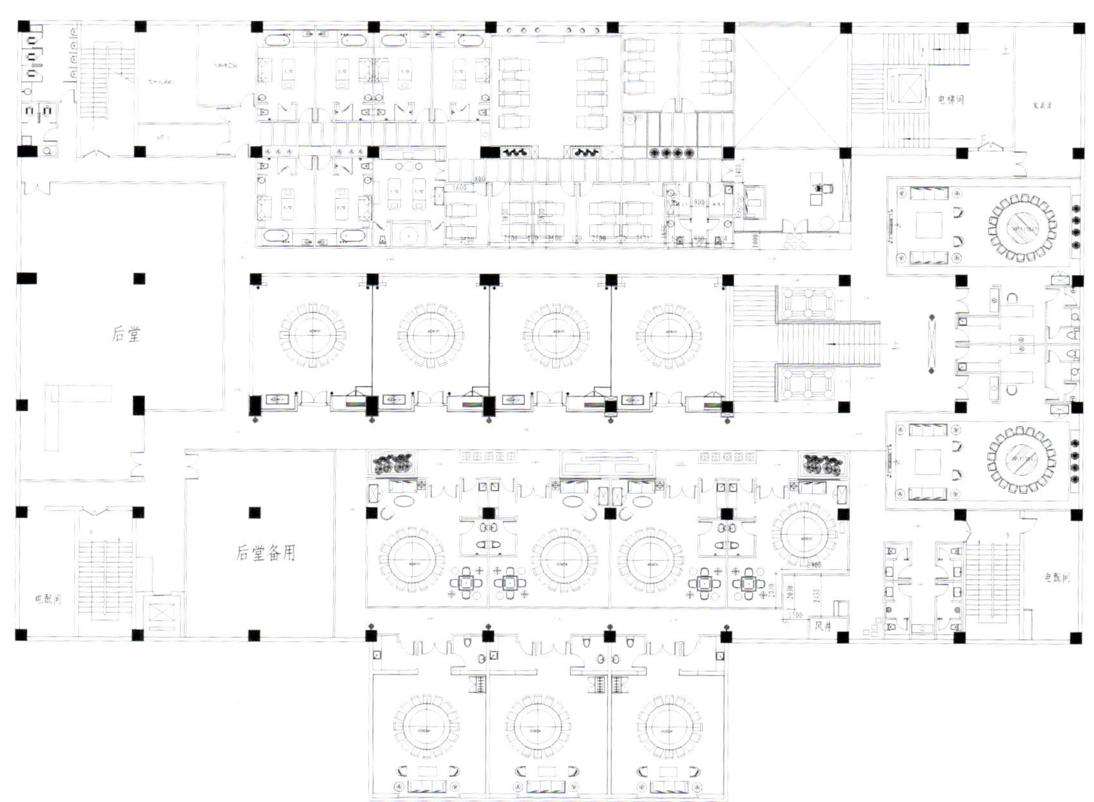
后堂
后堂备用

古城寿州

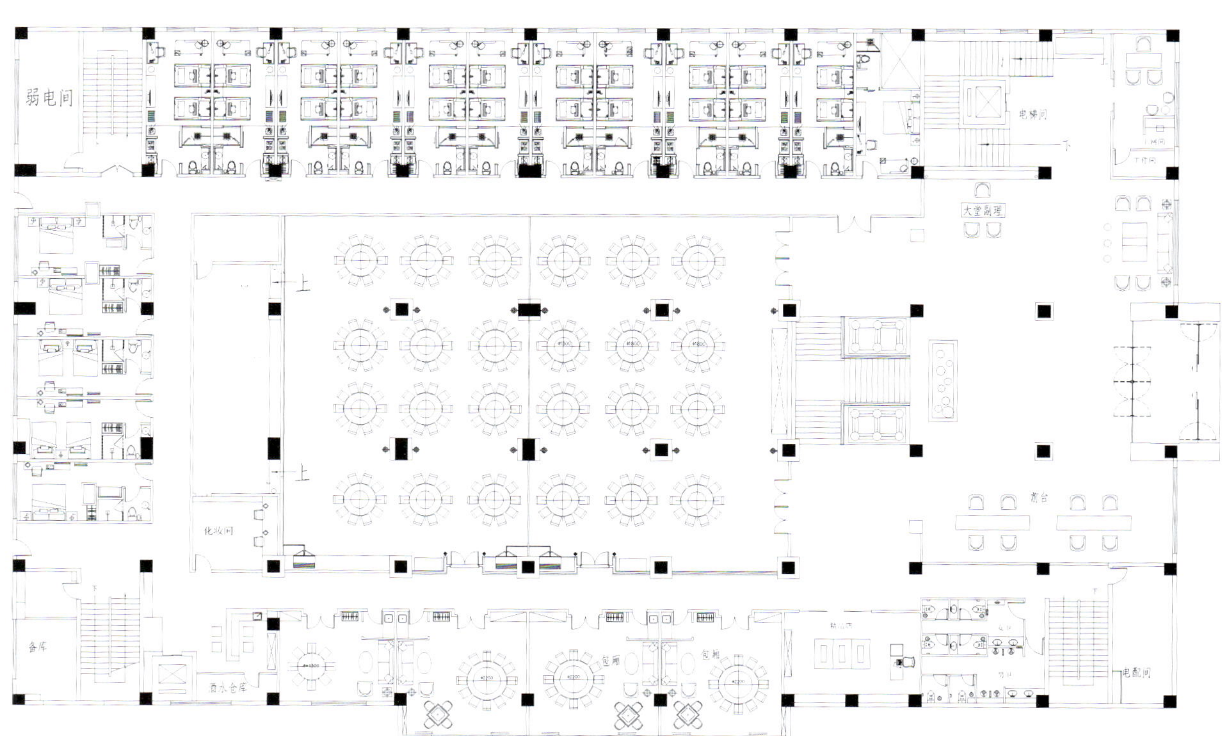
弱电间

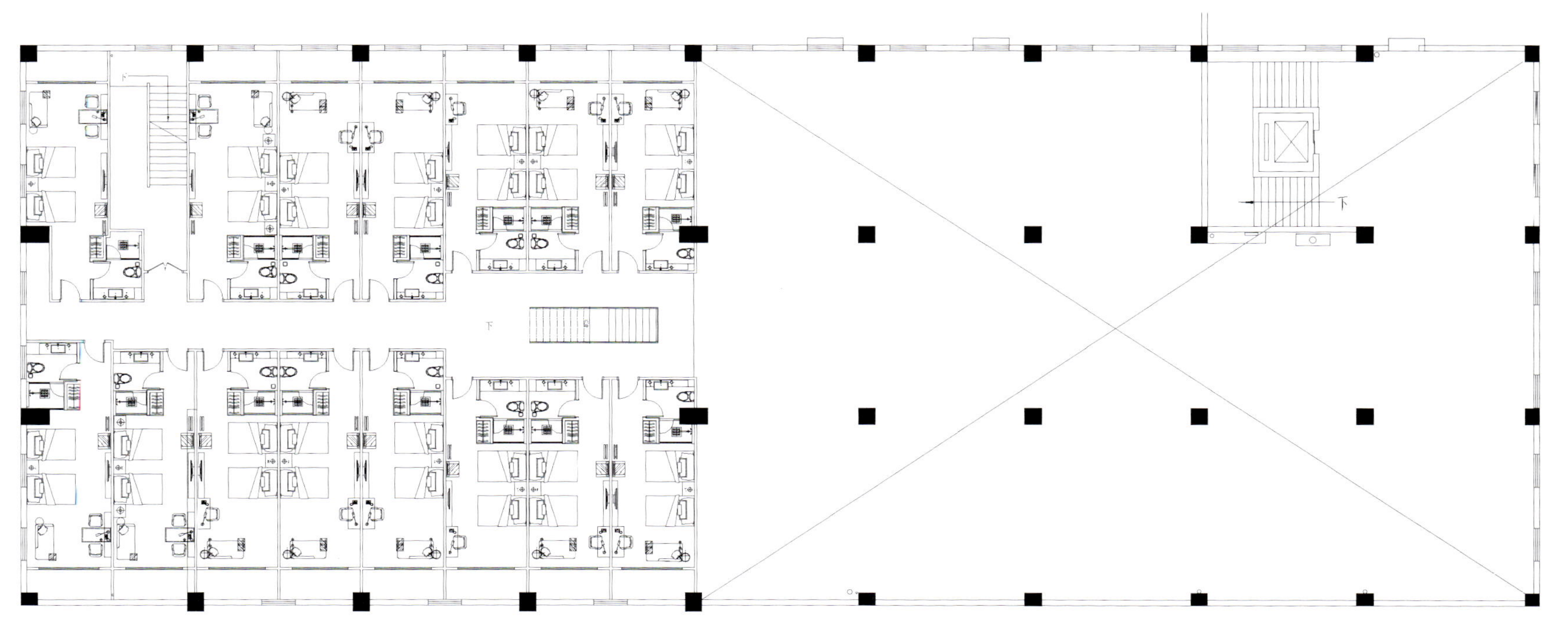

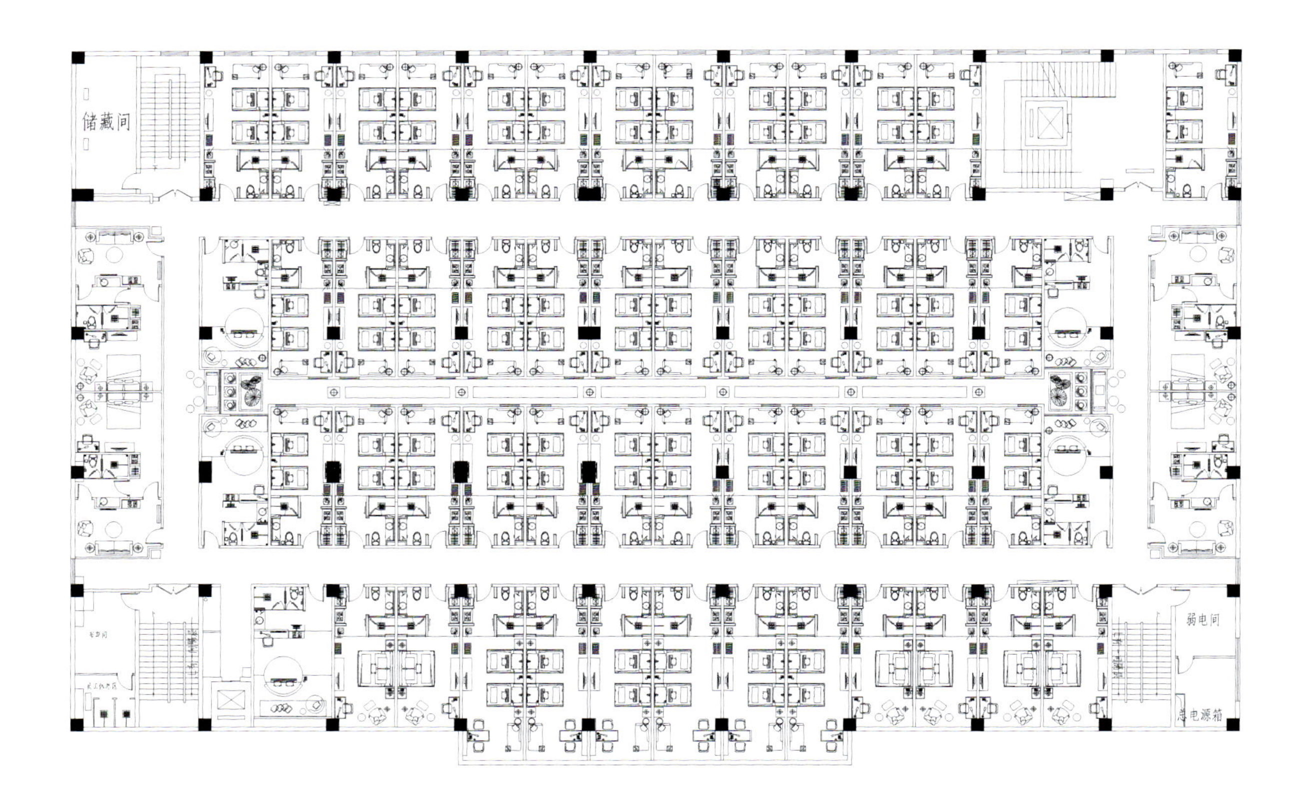

储藏间
弱电间
总电源箱

北京寿州大饭店以中式风格为设计主线，旨在传达中式情怀，展现中式文化底蕴和艺术魅力。项目面积较大，场地开阔，能充分体现中式风格的独特韵味和气质；加上地域文化的烘托，与周边环境很好的融合，更体现出了“风清木暖，荷香满溢”的情致。

饭店总体以中式风格为主，在客房的设计中融入其他风格，加入异域元素和饰品，充分体现了空间的包容性和艺术性。饭店功能分区明确而清晰，很好地平衡了空间感。在设计装饰方面，充分体现了中式建筑的特色与韵味，青砖、灰瓦、仿古砖、仿古家具、屏风、木质搁架等结合现代材料，打造出一个内敛、优雅的框架，再以经典的中式元素，如陶瓷、字画、木雕、石刻、竹丛等点缀填充，让空间更丰满，让生活更圆满。

8310

8269

8215
8213
8211
8209
8207
8220
8218

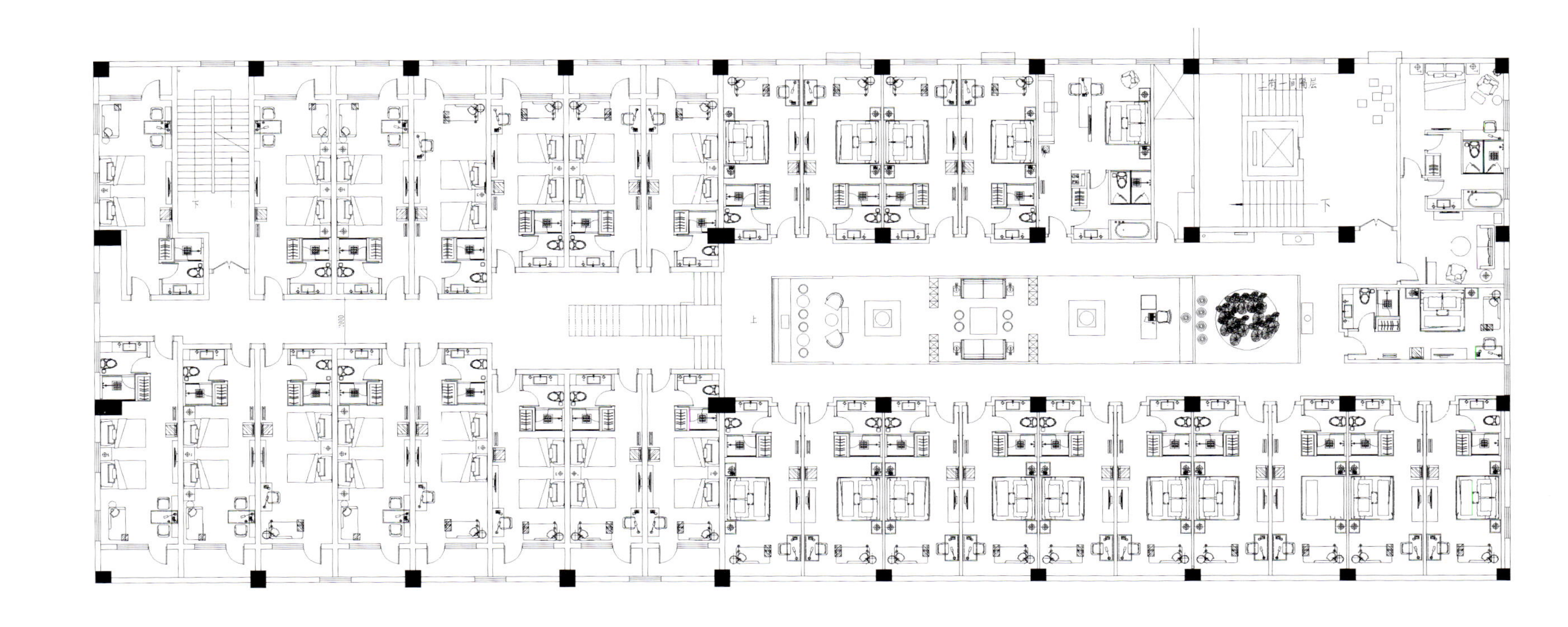

Kardier

Butterfly Therapeutic Retreat

项目地址
合肥

面积
460 平方米

设计师
许建国

参与设计
陈涛、欧阳坤、陈迎亚

设计公司
合肥许建国建筑室内装饰设计有限公司

摄影
吴辉

意兰庭休闲会所

Butterfly, the word itself is an elegant name, and the meaning of "butterfly" is reflected in the space, so the whole space is immersed in a quiet atmosphere, full of Zen mood. It features a pure land in the bustling city, allowing guests to put off the impetuous and tired feelings and to find their missing treasures.

Layers of blue bricks are neatly outlined; large areas of bluestone boards are connected between walls, featuring a mood of "raining dusk". Pieces of gray tiles are closely linked to bring guests back to the remote past, and their hearts are filled with inexplicable kindness and melancholy. Strips of wooden battens with scars are primitive and rustic, so guests will automatically think about the sentence that gentlemen are more plain than sophisticated. Pieces of lotuses open or ready to burst, and the delicate and shy lotuses are resisting cold, easily stirring the visitors' heartstrings. The stone block, the jar, and the touch of red and green...in the plain space they will bring guests a peace of mind, with satisfaction and happiness.

BUTTERFLY
BUTTERFLY
会所

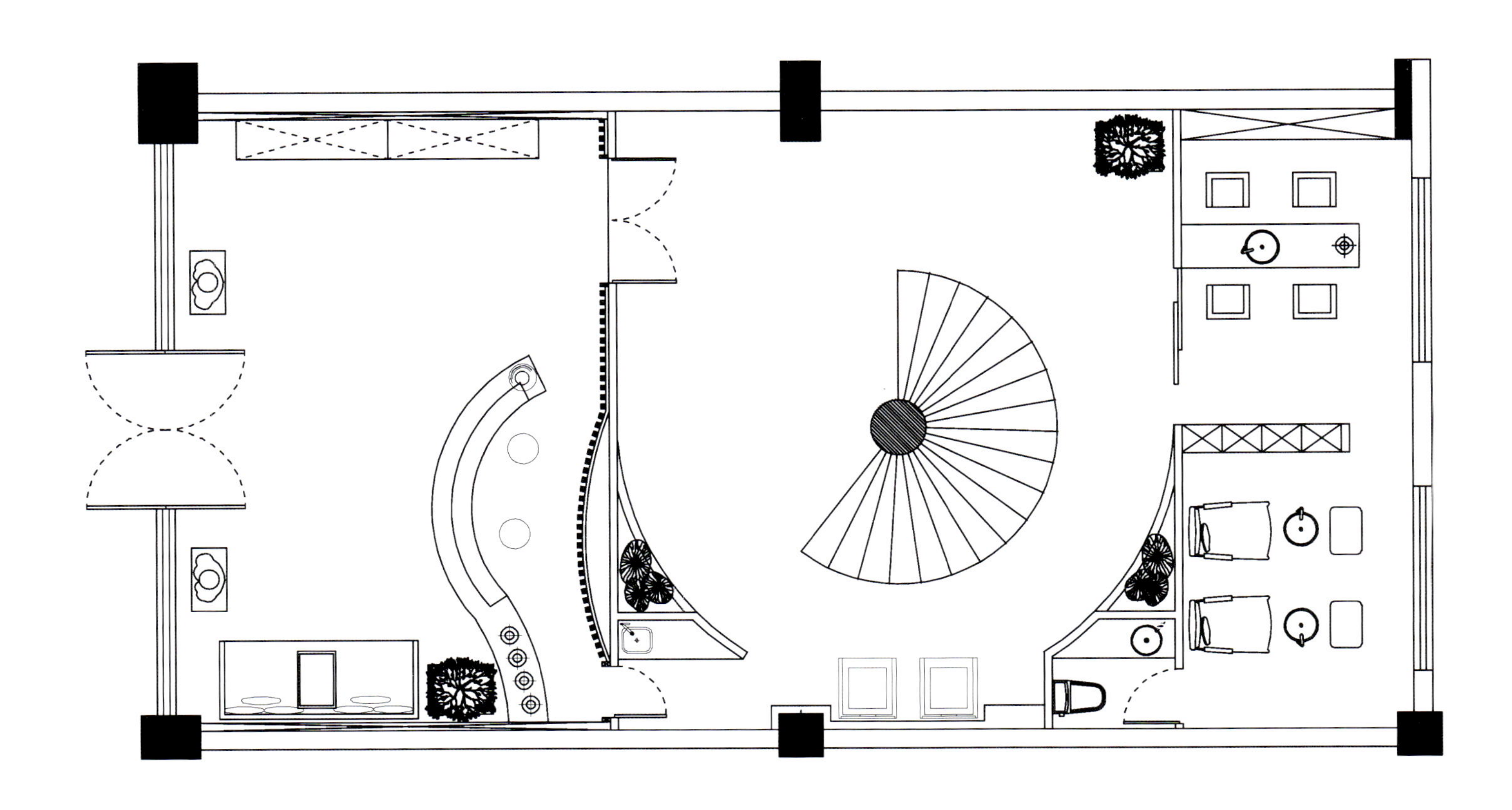

意兰庭，本身就是一个极雅致的名字，又有着“意兰亭”的含义在其中，于是整个空间便沉浸在一片静谧中，极富禅意。在喧嚣中辟出一方净土，让来客褪去一身的浮躁和疲惫，在这里找到遗失的珍宝。

一层一层整齐勾勒的青色的砖，在墙与墙之间，大片连着的青石板，带给人“烟雨黄昏”的意境。一片一片紧密连着的灰色的瓦，带人回到那久远的年代，让人心底涌起莫名的亲切和惆怅。一条一条带着疤节的木条，原始而拙朴，让人自然而然地想起来那句话：君子与其练达，不若朴鲁。还有一朵一朵或开或含的莲，娇羞的、不胜凉风的莲，轻易拨动来客的心弦；还有那石墩、那大缸，还有那一点红、那一抹绿……就在这平平淡淡中，给人带来心灵的宁静，带来满足，带来幸福。

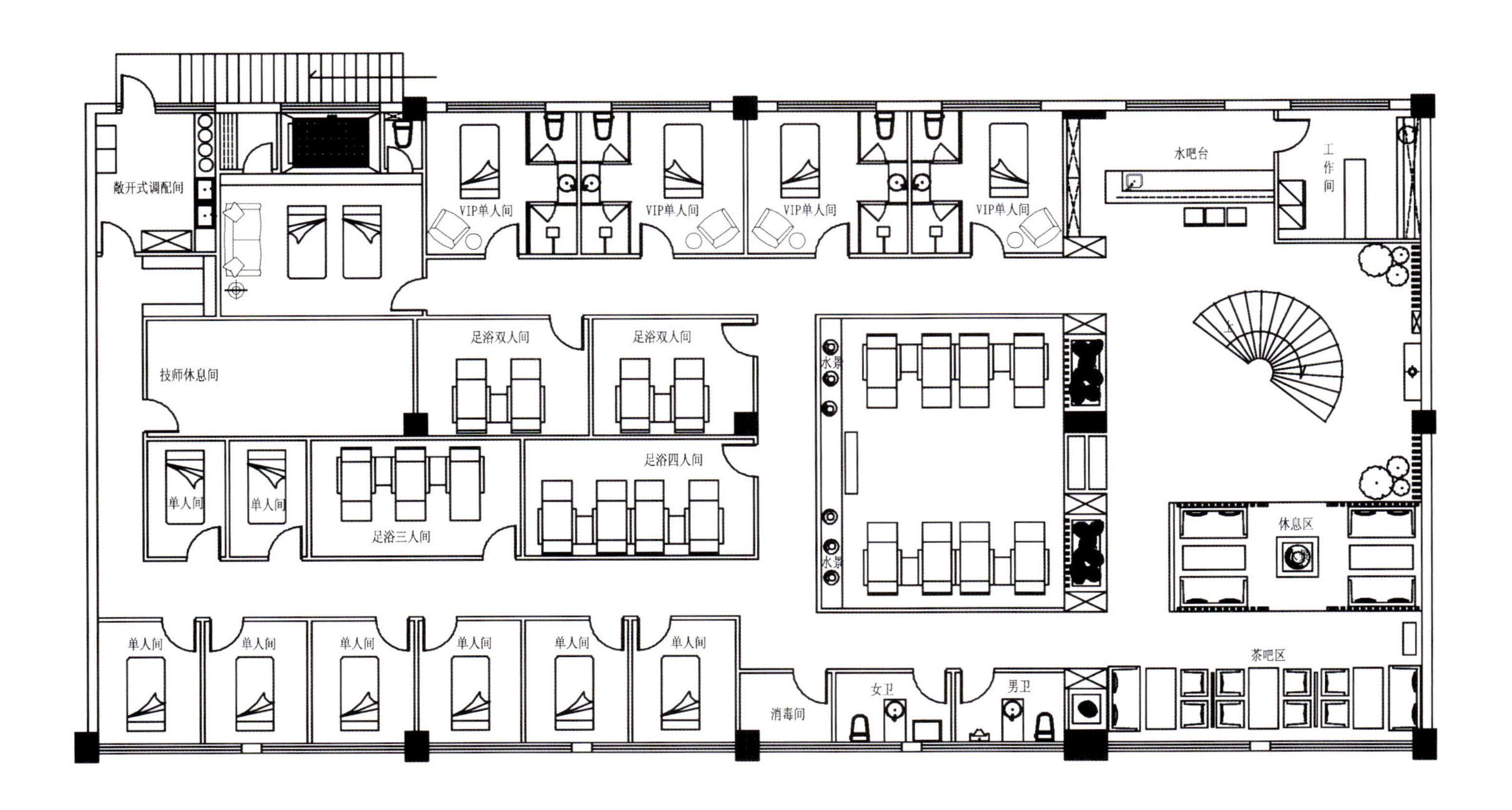

Office of Jianguo Xu Architectural Interior Design Limited Company

项目地址
安徽合肥

面积
130 平方米

设计师
许建国

设计公司
合肥许建国建筑室内装饰设计有限公司

参与设计
陈涛、欧阳坤、程迎亚

许建国建筑室内装饰设计有限公司办公室

摄影
吴辉

主要用材
旧房梁、旧木板、旧窗户、红砖、乳胶漆

The project is the designer's own studio, with focus on creating a relaxing and natural working environment; he puts all of his own preferences, thoughts and ideas into the space. There are simple furniture of modern style and charming Ming-dynasty furniture, and a lot of special objects collected by the designer – rotten iron sheets from abandoned garage, wooden doors removed from old houses, stone manger on roadside...all of them become parts of the space. The old objects are fully integrated with modern ones, making the old valueless stuff in other's eyes back to life, and endowing the studio with a distinctive taste.

"A design must have a story, it is not all about decorating." Here, the camera accompanying the designer for many years, the old woodcarving, the lamps personally designed by the designer –"angry birds", all of them are telling their own stories. Comfort, functionality and ornamental value of the space are perfectly integrated, so people who working in the office will feel relaxed and comfortable even in busy work.

本案是设计师本人的工作室，着重营造轻松自然的工作环境，并将自己的喜好、思想、理念全部灌注在这个空间中。这里有现代风格的简约家具和古韵十足的明式家具，还有许多设计师搜罗来的特别物件——从废弃修车厂找来的烂铁皮，旧房子上拆下来的木门，路边淘来的石头马槽……都成了空间的一部分。旧物与现代器物充分结合，既让这些别人眼中没有价值的东西重获生机，又为工作室增添了与众不同的趣味。

“设计一定要有故事，不能只有装饰。”在这里，跟随设计师多年的相机，度过了漫长光阴的老木雕，设计师亲自设计的灯具——“愤怒的小鸟”，都在讲述着自己的故事。空间的舒适性、功能性和观赏性得到了完美的结合，使得在此办公的人在忙碌的工作中也能寻找到放松与舒适。

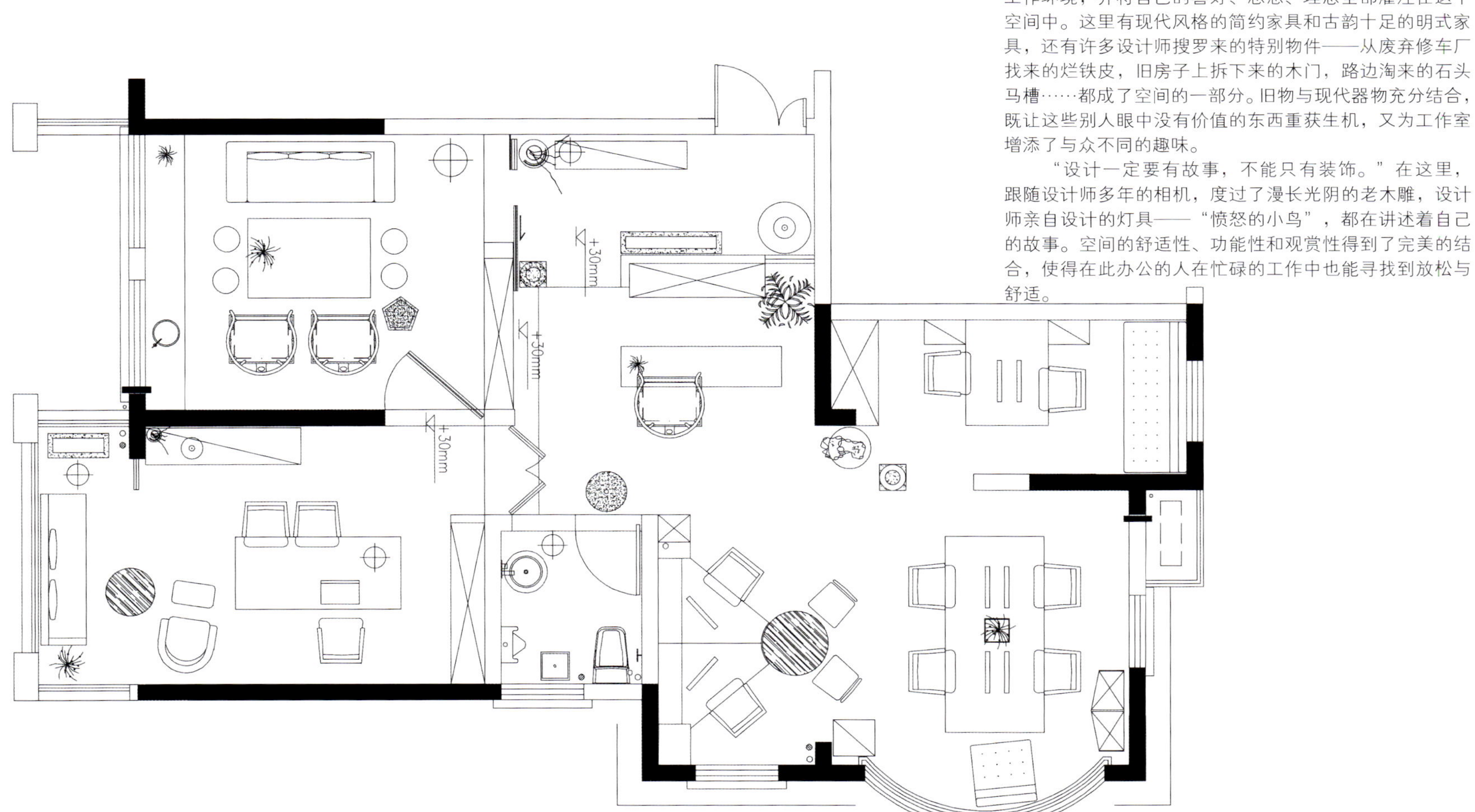

DORIS CHUI
徐 少 娴

Gotomaikan International Limited 设计总监

毕业于香港大学，之后深造于牛津大学东方研究所，精通中文、英文、日文、法文
1990 年 受聘于 Hirsch Bendner Associates 设计顾问有限公司
1996 年 在东京创立五斗米馆有限会社
1997 年 成立香港五斗米馆国际有限公司
2003 年 成立五斗米馆国际有限公司上海办事处

主要完成的室内设计及艺术顾问项目：
北京国际俱乐部饭店、天津泰达万丽酒店、上海瑞吉红塔大酒店、沈阳丽都喜来登酒店、重庆 J.W. 万豪酒店、北京国航万丽酒店、上海淳大万丽酒店、香港万豪酒店总统套房改造工程、上海浦东锦江索非特大酒店、上海兴国宾馆、北京棕榈泉国际公寓、珠江壹千栋千万量级豪华别墅等项目

Poly Pazhou Bay Lobby

项目地址
广州

面积
1533 平方米

设计师
徐少娴

设计公司
Gotomaikan International Limited

保利琶洲湾大堂

Poly Pazhou Bay is the first city's mixed-use project with floor area of millions of square meters in Guangzhou. It is comprised of five-star hotels, supper Grade A office buildings, river view mansions, high-grade apartments, large-sized shopping mall and international style commercial street. Therefore, the design of the lobby takes multi-function and humanization as the designing concept, aiming to create a gorgeous and noble, generous and dignified space. So clients can enjoy the scenery, and the building itself can also play proper business functions.

The whole design pays attention to the interaction and symmetry, whether in color or in shape, it is aimed at highlighting a generous, stylish and elegant space, putting the international style and sense of fashion into full play. The designer takes generous and calm neutral colors as the main tone, coupled with purple, gold, silver and other colors to highlight the noble and luxury sense, and the reflection of mirror, glass and metal materials and the effect of lighting equipments work together to set off an extraordinary gorgeous temperament. You will not feel raffish here; it shows the "magnificence" of a large space perfectly.

保利琶洲湾是集五星级酒店、超甲级写字楼群、江景豪宅、高级公寓、大型购物中心及国际风情商业街为一体的广州首个百万平方米城市综合体项目。因此，其大堂的设计以多元化、人性化为设计理念，意在打造一个华丽高贵、大气端方的空间，既能观能赏，又能发挥应有的商业功能。

整体设计讲究呼应、对称，不管是色彩方面，还是造型上，都以凸显大气、时尚、高雅为宗旨，将国际化风情和时尚感表现得淋漓尽致。设计师以大气沉稳的中性色为主调，加入紫、金、银等凸显高贵、奢华的色彩，并运用镜面、玻璃、金属材料的反射和照明设备的晕染，衬托华丽不凡的气质，既不会让人觉得艳俗，又很好地展示了大空间的“豪气”。

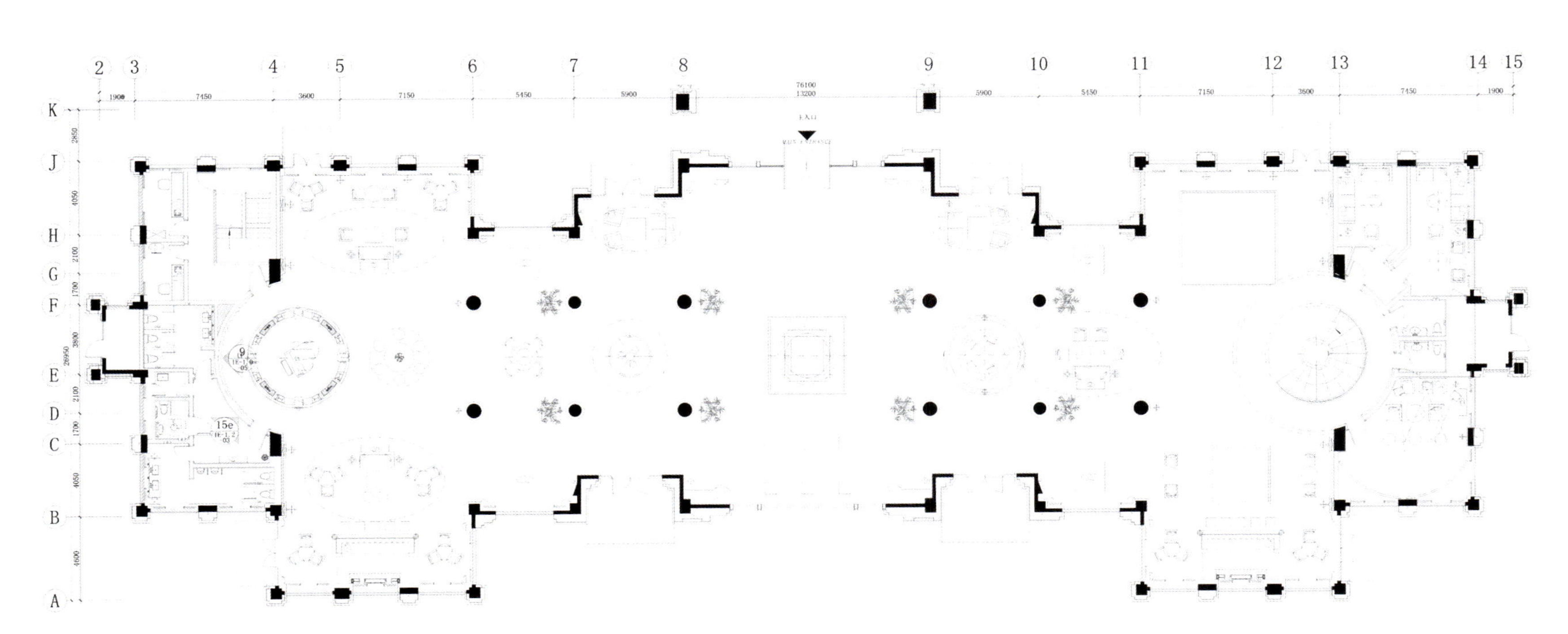

Hangzhou Shuguang Unit A

项目地址
杭州

面积
121.85 平方米

设计师
徐少娴

设计公司
Gotomaikan International Limited

杭州曙光 A 户型

The project focuses on the soft decoration and light rigid pavement as the designing principles, striving to create an outside gorgeous and inside warm home space, to provide a place for the owner to cultivate his body and mind and to enjoy the fun at home.

The pattern of three bedrooms, one living room and one dining room meets the owner's needs of living, and it takes a full and rational use of the space. The designer introduces neo-classical style to interpret the space, with the main tone of light colors to lay an elegant, warm and romantic tone for the space, and the light gold and bright silver are introduced to render and set off the space and to dress it with a layer of luxurious and elegant coat, fully showing the neo-classical flavor and charm.

In the space design, the designer creates the space layout with tough and concise lines, but in terms of soft decoration, she chooses furniture and ornaments with soft lines and elegant style to balance the sense of space, and to show the other taste of neo-classical style, fully reflecting the elegance and warm.

本案以注重软装饰、轻硬质铺装为设计原则，力图打造一个华丽其外，温馨在内的家居空间，为屋主提供一个休养身心、乐享天伦的场所。

三房两厅的格局既满足了屋主家庭生活的功能所需，又充分合理地利用了空间。设计师以新古典风格来诠释空间，以浅色系作为主色调，为空间奠定了雅致、温馨、浪漫的基调，并加入浅金色、亮银色来渲染、烘托，为空间镀上了一层奢华而优雅的外衣，充分体现了新古典的风情与魅力。

在空间设计上，设计师以硬朗简约的线条打造空间平面，而在软装修饰上，则选择线条柔和、造型优雅的家具、饰品来点缀，平衡空间感之余，将新古典风格的另一面展现出来，尽显优雅、温馨。

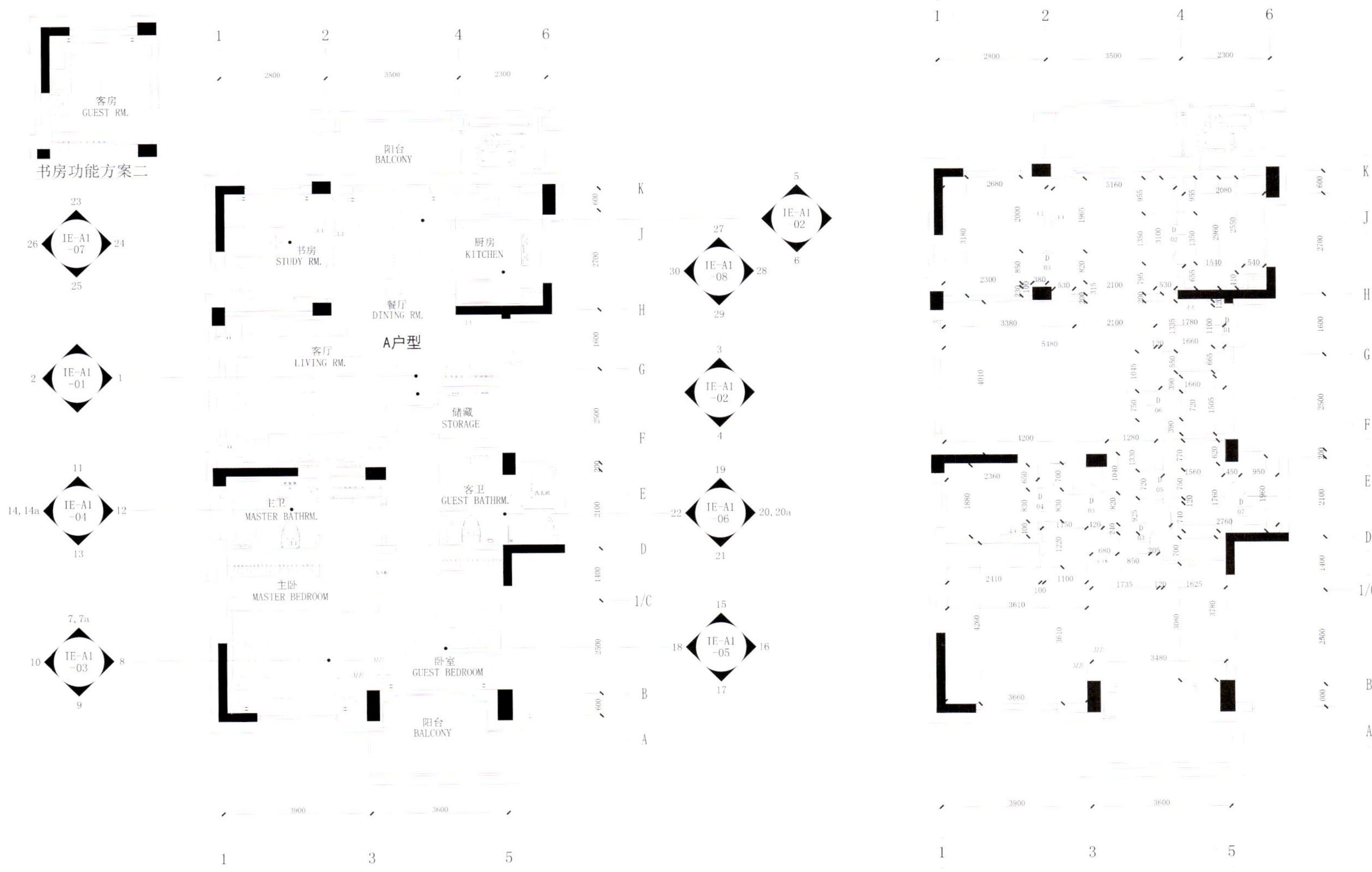

客房
GUEST RM.
书房功能方案二
阳台
BALCONY
书房
STUDY RM.
厨房
KITCHEN
餐厅
DINING RM.
A户型
客厅
LIVING RM.
储藏
STORAGE
主卫
MASTER BATHRM.
客卫
GUEST BATHRM.
主卧
MASTER BEDROOM
卧室
GUEST BEDROOM
阳台
BALCONY

Hangzhou Shuguang Unit C

项目地址
杭州

面积
112 平方米

设计师
徐少娴

设计公司
Gotomaikan International Limited

杭州曙光 C 户型

The neo-classical style features classicism and modernism, perfectly showing an elegant, fashionable and romantic style. In the design, the project integrates neo-classical style, fully reflecting the function of "home", showing the charm of the designing style, to make preparation to show the owner's taste and personality.

It is mainly pure and peaceful white, coupled with colors of black, brown, light gold and bright silver, in balanced proportion and appropriate shade, to show a simple and noble space. In the aspect of furniture and furnishings, furniture focuses on reflecting the principle of comfort, and furnishings play a role of finishing touch, just in the right amount, lightening up the space instantly and setting off the space atmosphere. Somehow it pays attention to the match of black and white, continuing the classic match of black and white.

Generally speaking, the space reveals a concise and fashionable temperament, and the neo-classical style inadvertently emerges, fully interpreting the flavor of "looming and specious".

新古典兼具古典主义与现代主义设计风格，能完美地体现典雅、时尚以及浪漫的风情。本案在设计中融入新古典风格，在充分体现“家”的功能的同时，也将设计风格的魅力一一展现，为体现屋主的品位和个性做出铺垫。

以纯净、安宁的白色为主，加入黑色、棕色、浅金、亮银，比例均衡，浓淡合宜，将一个简约而尊贵的空间展示出来。在家具陈设方面，家具重点体现舒适性的原则，陈设则起着画龙点睛的效果，不多不少，却能在瞬间点亮空间，烘托出空间氛围，并在一定程度上讲究黑白搭配，续写黑白配的经典。

总体来说，空间更多地流露出简约、时尚的气质，而新古典的风情却在不经意间崭露头角，将“若隐若现、似是而非”的风情诠释得淋漓尽致。

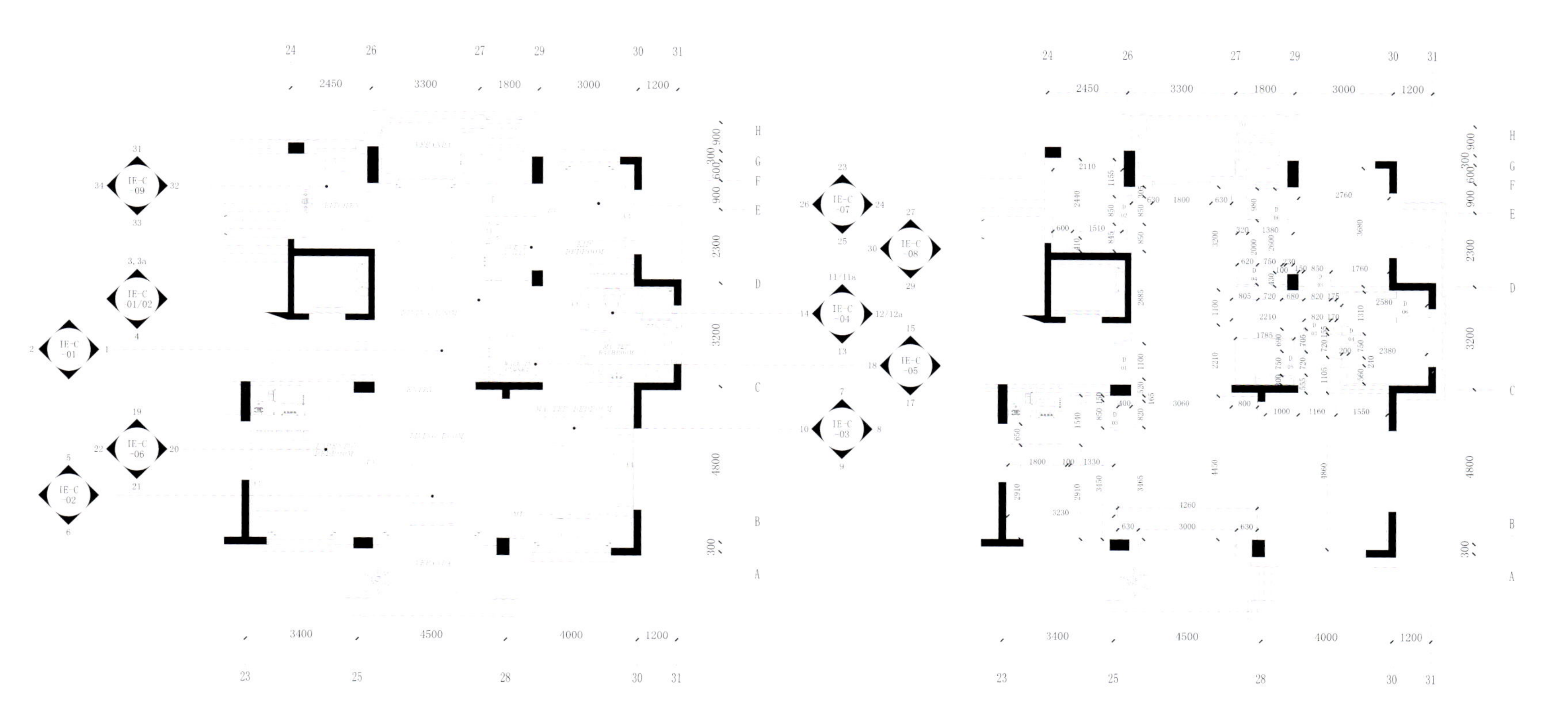

Hangzhou Shuguang Unit D

项目地址
杭州

面积
70 平方米

设计师
徐少娴

设计公司
Gotomaikan International Limited

杭州曙光 D 户型

The floor area of this project is small, but it meets the function of three bedrooms, one living room and one dining room, truly reflecting the meaning of "small but perfectly formed".
The designer takes modern fashion as the designing idea, fully reflecting the essence of fashion in color application and furniture styling, to create a home space with practical functions and elegant appearance for the owner. In the use of colors, the match of black and white is the main tone, embellished with brown, blue and red to highlight the concise and fashionable theme. In the selection of furniture, in addition to adhering to the principle of comfort, the designer also chooses some unique furniture with elegant styles and gorgeous colors as the finishing touch, coupled with exquisite and fashionable lightings and modern crafts to fully show the personality and taste of a small fashionable home, meanwhile to highlight the preference and style of the owner.

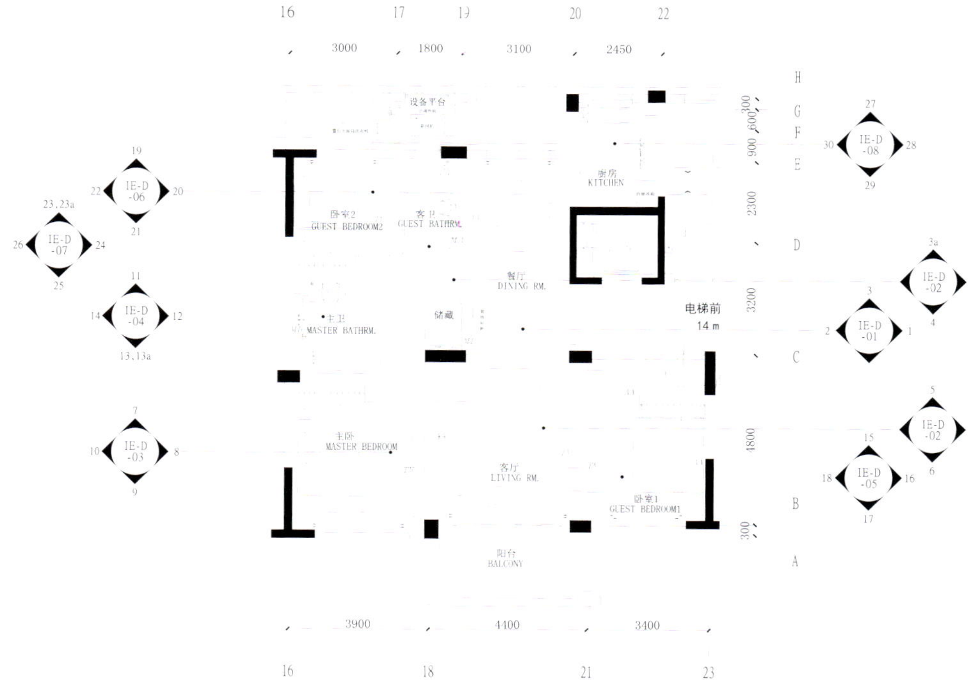

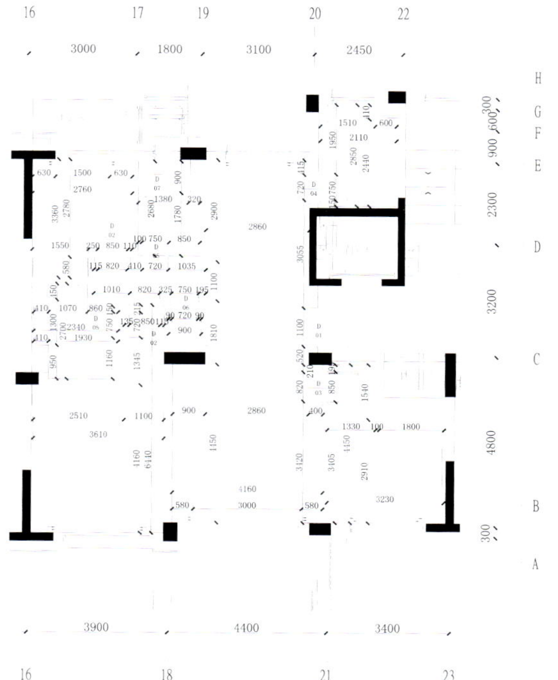

本案属于小户型项目，但是空间同样做足了三房两厅的功能，真正体现了“麻雀虽小，五脏俱全”的含义。

设计师以现代时尚为设计思路，在色彩运用和家具造型方面充分体现时尚这一要领，为屋主打造了一个功能实用而外形优雅的家居空间。在色彩运用上，以黑白搭配为主色调，加入棕色、蓝色和红色作为点缀，凸显了简约时尚的主题。在家具的选择上，除了坚持舒适原则，设计师还选择了一些造型优雅、色彩艳丽的个性家具作为点睛之笔，再配上精巧、时尚的灯饰和现代工艺品，充分展现出时尚小家的个性与品位，同时也彰显了屋主的喜好与格调。